Beyond Michaelis-Menten: Modernizing the Equation for Enzyme Kinetics

Nicholas

Contents

Introduction

I. The Michaelis-Menten Equation

The year of 2013 marked the one hundredth anniversary of the publication of the classic Michaelis-Menten (MM) paper 'Die Kinetik der Invertinwirkung' [1] in which the MM equation was first proposed. This work depicted the study of invertase (beta-fructofuranosidade, EC 3.2.1.26), an enzyme that catalyzes the conversion of sucrose to fructose and glucose, which induces an inversion of optical rotation from positive (for the substrate) to an overall negative value (for the mixture of products) [2]. In its original portrayal, the MM equation expressed the rate v of the reaction as a function of the concentration of substrate $[S]$ (sucrose), with Φ as the total molar concentration of enzyme (invertase), k as the dissociation constant of the enzyme-substrate complex (an equilibrium constant and not a rate constant, despite being originally represented as a lower-case character), and C as a constant of proportionality.

$$v = C\Phi \frac{[S]}{[S] + k} \tag{I.1}$$

In modern formality, $C\Phi$ corresponds to V, the limiting rate for a given enzyme concentration, C corresponds to the catalytic constant k_{cat} (also known as turnover number), which defines the number of catalytic cycles that an enzyme can perform per unit time [3], and Φ is the total molar concentration of enzyme; k corresponds to K_S, the equilibrium dissociation constant of the enzyme-substrate complex.

Built on the work of earlier authors such as Adrian Brown [4] and Victor Henri [5,6], the MM methodology became the standard approach to steady-state enzyme kinetics. Michaelis and Menten understood the significance of pH control in enzymatic experiments and acknowledged that initial rates were easier to interpret than time courses because the latter are more affected by the reverse reaction, product inhibition and enzyme inactivation [3].

I.I. Briggs-Haldane Reaction Scheme

Modern representations of the MM equation employ the reaction scheme described in 1925 by Briggs and Haldane characterizing the reversible formation of an enzyme-substrate complex followed by its irreversible transformation into product and release of free enzyme:

$$E + S \underset{k_{-1}}{\overset{k_1}{\rightleftharpoons}} ES \overset{k_2}{\rightarrow} E + P \tag{I.2}$$

where k_1 and k_{-1} are the rate constants associated with the reversible binding step, and k_2 is the rate constant corresponding to the catalytic step [7]. The concentrations of the different species change with time t as described by the following system of first-order differential equations:

$$\frac{d[S]}{dt} = -k_1[E][S] + k_{-1}[ES] \tag{I.3}$$

$$\frac{d[ES]}{dt} = k_1[E][S] - (k_{-1} + k_2)[ES] \tag{I.4}$$

$$\frac{d[E]}{dt} = -k_1[E][S] + k_{-1}[ES] + k_2[E] \tag{I.5}$$

$$\frac{d[P]}{dt} = k_2[ES] \tag{I.6}$$

subject to the initial conditions $([S],[E],[ES],[P]) = (S_0, E_0, 0,0)$. Although the analytical solution of Eqs. I.3-I.6 is not known [8], this system of equations can be simplified by application of the steady-state approximation (SSA). According to this approximation, the concentration of the enzyme-substrate ($[ES]$) complex remains constant in the presence of a large excess of substrate once the initial transient period has elapsed [7,9]. Considering the following mass conservation laws,

$$E_0 = [E] + [ES] \tag{I.7}$$

$$S_0 = [S] + [ES] + [P] \tag{I.8}$$

and the abovementioned SSA approximation,

$$\frac{d[ES]}{dt} = 0 \tag{I.9}$$

it is possible to simplify the system of ordinary differential equations composed by Eqs. I.3-I.6 and obtain the expression of the reaction rate $v = k_2[ES]$ as a function of the substrate concentration [7]:

$$v = k_2[ES] \Leftrightarrow v = \frac{k_2 E_0[S]}{\frac{k_{-1} + k_2}{k_1} + [S]} \tag{I.10}$$

Eq. I.10 can be written in a more general form describing the hyperbolic dependence of the initial reaction rate on the substrate concentration [3]:

$$v = \frac{k_{cat}E_0[S]}{K_m + [S]} = \frac{V[S]}{K_m + [S]} \tag{I.11}$$

In addition, if the duration of the transient initial period is short enough to assume invariant $[S] \approx S_0$, the Reactant Stationary Approximation (RSA) is applicable, and the final form of the MM equation is obtained:

$$v_0 = \frac{VS_0}{K_m + S_0} \tag{I.12}$$

where v_0 is the initial reaction rate, V the limiting rate and K_m the Michaelis constant. In the Briggs-Haldane notation, $V = k_2E_0$, while $K_m = (k_{-1} + k_2)/k_1$. In order to extend the use of the rate constant k_2 to more complex reaction schemes, V is written as $k_{cat}E_0$, where the turnover number k_{cat} can represent more than one elementary step.

II. PEA Model

Although the MM equation (Eq. I.12) presents a useful simplification of the system of equations I.3-I.6, its validity is restricted to the initial phases of enzymatic reactions with great excess of substrate over the enzyme [10-12]. A vast and important region of conditions is, therefore, ignored when using the MM equation, especially when other timescales than the initial moments of the enzymatic reaction are considered. The publication of the Pinto *et al.* (PEA) model in 2015 [10] contributed to reveal the "whole picture" of single active-site enzyme kinetics without inhibition [10]. More specifically, considering $S_0 \gg E_0$ the 'white' region for which the MM equation is valid, the complementary 'gray' ($S_0 \sim E_0$) and 'dark' ($S_0 \ll E_0$) regions were uncovered for all timescales of enzyme catalysis [10].

II.I. Formulation

As mentioned above, the system of first-order differential equations describing the Briggs-Haldane reaction scheme (Eqs. I.3-I.6) does not have a known analytical solution [8]. However, by selecting the accessible pivotal variable $(S_0 - P)/v$, in which $(S_0 - P)$ is the concentration of product still to be formed, a closed-form solution of this system can be derived [10]. The pivotal variable represents how much time would be required for reaction completion if the instantaneous rate of reaction was kept constant. Although an approximate solution, the PEA model is valid for every combination of model parameters and variables, thereby surpassing the restrictions imposed by the MM equation [10].

The system of first-order differential equations described by Eqs. I.3-I.6 constitutes the starting point for the derivation of the PEA model. It is possible to eliminate the equation corresponding to $d[E]/dt$ from this system by employing the mass conservation law presented in Eq. I.7:

$$\frac{d[S]}{dt} = -k_1(E_0 - [ES])[S] + k_{-1}[ES] \tag{I.13}$$

$$\frac{d[ES]}{dt} = k_1(E_0 - [ES])[S] - (k_{-1} + k_2)[ES] \tag{I.14}$$

$$\frac{d[P]}{dt} = k_2[ES] \tag{I.15}$$

The concentration of the different species can be normalized by K_m as $s = [S]/K_m$, $e_0 = [E_0]/K_m$, $c = [ES]/K_m$, $p = [P]/K_m$, with $K_S = k_{-1}/k_1$; Eqs. I.13-I.15 can also be expressed as a function of the scaled time $\theta = k_2 t$:

$$-\left(1 - \frac{K_S}{K_m}\right)\frac{ds}{d\theta} = e_0 s - c\left(\frac{K_S}{K_m} + s\right) \tag{I.16}$$

$$\left(1 - \frac{K_S}{K_m}\right)\frac{dc}{d\theta} = e_0 s - c(1 + s) \tag{I.17}$$

$$\frac{dp}{d\theta} = c \tag{I.18}$$

The Supplementary Information provided by Pinto *et al.* (2015) [10] describes in detail the different steps and the three approximations required to derive the PEA model from Eqs. I.16-I.18. The obtained analytical solution is represented as a function of s_0, e_0, and the scaled variables $\tau = K_m/V$, $\theta = t/(e_0\tau)$ and $\beta = 1 - K_S/K_m$.

A simplified form of the final solution is given by Eq. I.19a, where the pivotal variable is defined in terms of the Lambert function. This function constitutes the inverse of the function $f(x) = xe^x$, corresponding to the solution to the transcendental equation $W(x)e^{W(x)} = x$, denoted as $W(x)$ [13]. This formulation departs from definitions valid for $S_0 \gg E_0$ ('white' region of conditions) and incorporates a time-dependent correction factor Φ_C (Eq. I.19b) that accounts for the influence of regions of lower s_0/e_0.

$$\frac{S_0 - P}{v} = \tau\left[1 + \omega\left(s_0 exp(s_0 - t/\tau)\right)\right]\Phi_C \tag{I.19a}$$

$$\Phi_C = \frac{1}{2}\left(\frac{1 + e_0 + \tilde{s} + \tilde{\lambda}\Big/ tanh\left(\frac{\tilde{\lambda}\theta}{2\beta}\right)}{1 + \omega\left(s_0 exp(s_0 - e_0\theta)\right)}\right) \tag{I.19b}$$

II.II. Estimation of kinetic parameters with the PEA Model

The stationary state of an enzymatic reaction is reached after the initial transient period of ES complex build-up has elapsed. The enzymatic reaction rate reaches its maximum value at the stationary instant t^* corresponding to the end of this build-up phase. The pivotal variable at this instant $((S_0 - P^*)/v^*)$ is independent of K_S, as demonstrated in the Supplementary Information provided by Pinto $et\ al.$ (2015) [10]. Hence, considering stationary-state conditions, an algebraic simpler form of Eq. I.19a is obtained (Eq. I.20). Its use allows for simple and universal determination of MM parameters, namely K_m and V, through the application of simple linear regressions and without major experimental limitations other than the instrumental resolution necessary to observe stationary moments.

$$\frac{S_0 - P^*}{v^*} = \frac{\tau}{2}\left(2 + s_0 + e_0 + |s_0 - e_0|\right) \Leftrightarrow \frac{S_0 - P^*}{v^*} = \begin{cases} \frac{K_m + S_0}{V}, & S_0 > E_0 \\ \frac{K_m + E_0}{V}, & S_0 < E_0 \end{cases} \tag{I.20}$$

The stationary state version of the PEA model describes the linear influence of E_0^{-1} over the stationary pivotal variable for $S_0 < E_0$ and fixed S_0, as well as the linear influence of S_0 over the same variable for $S_0 > E_0$ and fixed E_0. Under non-MM conditions the highest reaction rates may not coincide with the initial reaction rates [10]. Therefore, Eq. I.20 eliminates the ambiguity associated to the use of instantaneous reaction rate methods. For maximal values of the dissociation constant ($K_S/K_m = 1$) and $S_0 > E_0$, Eq. I.20 is reduced to the MM equation, while for $S_0 < E_0$ it reduces to the simplified Bajzer and Strehler equation [14] (Eq. I.21):

$$v_0 = \begin{cases} \dfrac{VS_0}{K_m + S_0}, & S_0 > E_0 \\ \dfrac{VS_0}{K_v + E_0}, & S_0 < E_0 \end{cases} \tag{I.21}$$

References

1. Michaelis, L. and Menten, M. (1913), *Die Kinetik der Invertinwirkung.* Biochemische Zeitschrift. 49:333–369.

2. Johnson, K.A. and Goody, R.S. (2011), *The Original Michaelis Constant: Translation of the 1913 Michaelis-Menten Paper.* Biochemistry. 50:8264-8269.

3. Cornish-Bowden, A., (2012), Fundamentals of Enzyme Kinetics, 4th ed. Wiley-Blackwell (Weinheim, Germany), Chp. 2.

4. Brown, A.J. (1902), *XXXVI.—Enzyme action.* Journal of the Chemical Society, Transactions. 81:373-388.

5. Henri, V. (1902), *Théorie générale de l'action des quelques diastases.* Comptes rendus hebdomadaires des séances de l'Académie des sciences. 135:916-919.

6. Henri, V., (1903), *Lois Générales de l'action des Diastases.* Librairie Scientifique A. Hermann (Paris).

7. Briggs, G.E. and Haldane, J.B.S. (1925), *A note on the kinetics of enzyme action.* Biochemical Journal. 19:338-339.

8. Berberan-Santos, M.N. (2010), *A General Treatment of Henri-Michaelis-Menten Enzyme Kinetics: Exact Series Solution and Approximate Analytical Solutions.* MATCH Communications in Mathematical and in Computer Chemistry. 63:283-318.

9. Cornish-Bowden, A. (2013), *The origins of enzyme kinetics.* FEBS Letters. 587:2725-2730.

10. Pinto, M.F., *et al.* (2015), *Enzyme kinetics: the whole picture reveals hidden meanings.* FEBS Journal. 282(12):2309-2316.

11. Segel, L.A. (1988), *On the validity of the steady state assumption of enzyme kinetics.* Bulletin of Mathematical Biology. 50(6):579-593.

12. Hanson, S.M. and Schnell, S. (2008), *Reactant Stationary Approximation in Enzyme Kinetics.* The Journal of Physical Chemistry A. 112:8654-8658.

13. Stewart, S. (2005), *A new elementary function for our curricula?* Australian Senior Mathematics Journal. 19(2).

14. Bajzer, Z. and Strehler, E.E. (2012), *About and beyond the Henri-Michaelis-Menten rate equation for single-substrate enzyme kinetics.* Biochemical and Biophysical Research Communications. 417(3):982-985.

Chapter 1.

In search of lost time constants and of non-Michaelis-Menten parameters

The content of the present chapter constitutes the first peer-reviewed article published in the context of the present book (Article I, see Annex 1):

Pinto, M.F. and Martins, P.M. (2016), In search of lost time constants and of non-Michaelis-Menten parameters. Perspectives in Science. 9:8-16. doi:10.1016/j.pisc.2016.03.024

1.2. <u>Introduction</u>

The year of 2013 marked the one hundredth anniversary of the publication of the classic MM paper *Die Kinetik der Invertinwirkung* [1], which became the standard approach to steady-state enzyme kinetics. Supported by the work of earlier authors, most notably Adrian Brown [2] and Victor Henri [3,4], MM understood the significance of pH control in enzymatic experiments and acknowledged that initial rates were easier to interpret than time courses as they are not restrained by issues such as the reverse reaction, product inhibition or enzyme inactivation [5]. Modern representations of the MM model use the Briggs-Haldane reaction scheme encompassing the reversible combination of free enzyme E and substrate S to form the enzyme-substrate complex ES followed by its irreversible transformation into product P and release of enzyme (Eq. 1.1) [6]:

where k_1 and k_{-1} are the rate constants of the reversible binding step and k_2 is the rate constant of the catalytic step. The evolution of the concentration of the different species with time t is mathematically described by the following system of first-order differential equations:

$$\frac{d[S]}{dt} = -k_1[E][S] + k_{-1}[ES]$$
(1.2)

$$\frac{d[ES]}{dt} = k_1[E][S] - (k_{-1} + k_2)[ES]$$
(1.3)

$$\frac{d[E]}{dt} = -k_1[E][S] + k_{-1}[ES] + k_2[E]$$
(1.4)

$$\frac{d[P]}{dt} = k_2[ES]$$
(1.5)

subject to the initial conditions $([S], [E], [ES], [P]) = (S_0, E_0, 0, 0)$. Although the analytical solution of Eqs. 1.2-1.5 is not known [7], a simplified alternative results from adopting the steady-state approximation (SSA) stating that, in the presence of a large excess of substrate, the concentration of the enzyme-substrate complex remains constant after the initial ES build-up period has ended [6]. If, in addition, the duration of the transient period is short enough to assume invariant $[S]$, the reactant stationary approximation (RSA) is applicable [8], and the final form of the MM equation is obtained (Eq. 1.6):

$$v_0 = \frac{VS_0}{K_m + S_0}$$
(1.6)

with v_0 being the initial reaction rate, V the limit reaction rate obtained for very high substrate concentration values, and K_m the Michaelis constant. In the Briggs-Haldane notation V corresponds to k_2E_0 and K_m corresponds to $(k_{-1} + k_2)/k_1$; in practical terms, V is written as $k_{cat}E_0$ in order to extend its use to reaction schemes of higher complexity than Briggs-Haldane, while K_m is commonly referred as the concentration of substrate for which $v_0 = 0.5V$. The SSA and the RSA approximations severely limit the applicability of the MM equation to the initial phases of enzymatic reactions that start with great substrate excess over the enzyme ($S_0 \gg E_0$) [8-10]. With the publication of the Pinto *et al.* (PEA) model in 2015 [9] (see Introduction, Section II), additional threats associated to the usage of the classical formalism were identified, at the same time that the "whole picture" of single active-site enzyme kinetics without inhibition was revealed [9]. The PEA model also uncovered new applications or "hidden meanings" in the Briggs-Haldane mechanism, of which the present contribution particularly focuses the cases of the characteristic time constant τ_∞ and of the dissociation constant K_S . These parameters were chosen as they help to answer some of the new problems posed by Systems Biology while studying increasingly realistic enzymatic networks. Not only that, the following sections illustrate how τ_∞ and K_S can be used to characterize enzymatic activity, enzymatic efficiency and enzyme-substrate affinity in a straightforward and unambiguous manner.

1.3. <u>Numerical Procedures</u>

The system of differential equations describing the Briggs-Haldane reaction scheme (Eqs. 1.2-1.5) was expressed in normalized units as Eqs. 1.7-1.9 [9]:

$$-\left(1 - \frac{K_S}{K_m}\right)\frac{ds}{d\theta} = e_0 s - c\left(\frac{K_S}{K_m} + s\right) \tag{1.7}$$

$$\left(1 - \frac{K_S}{K_m}\right)\frac{dc}{d\theta} = e_0 s - c(1 + s) \tag{1.8}$$

$$\frac{dp}{d\theta} = c \tag{1.9}$$

where $s = [S]/K_m$, $c = [ES]/K_m$, $p = [P]/K_m$, $e_0 = [E_0]/K_m$, $\theta = k_2 t$ and $K_S = k_{-1}/k_1$. Enzymatic reaction progress curves showing the evolution of scaled product concentration p over scaled time θ were simulated with Mathworks® MATLAB R2013b. A script was developed to this end in which a MATLAB ordinary differential equation (ODE) solver was employed to numerically solve Eqs. 1.7-1.9 over the scaled time. The specific ODE solver used to this effect was *ode45*, a one-step solver (i.e. when computing the solution for t_n, the solver only requires the solution at the immediately preceding time point, t_{n-1}) based on an explicit Runge-Kutta (4,5) formula, the Dormand-Prince pair [11]. Numerical solutions were obtained over different ranges of integration of θ for limiting values of the scaled dissociation constant K_S/K_m and for different sets of e_0 and s_0 initial conditions.

1.4. <u>Results</u>

1.4.1. *The Characteristic Time Constant (τ_∞) and the Enzyme Efficiency*

The analytical solution describing single active-site enzyme kinetics without inhibition was obtained after introducing the "pivotal variable" $(S_0 - P)/v$ representing the concentration of product still to be formed $(S_0 - P)$ over the instant reaction rate v [9]. Figure 1.1A illustrates the physical meaning of the pivotal variable as the period of time τ_n that would be required to complete the reaction if the instant reaction rate was maintained. Alternatively, the negative reciprocal of this variable is promptly computed as the instantaneous slope of the $(S_0 - P)$ time-course curve represented in a log-linear scale (Figure 1.1B).

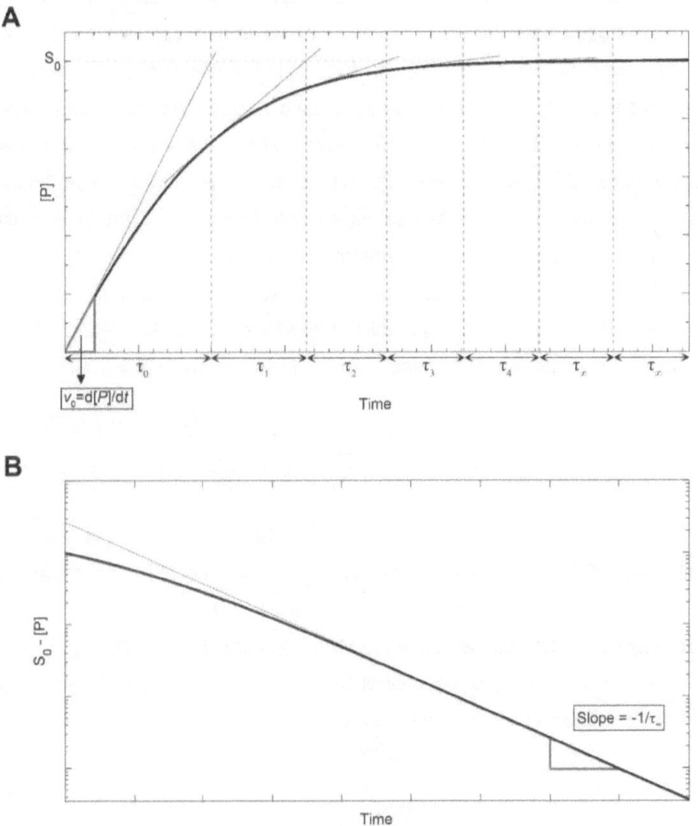

Figure 1.1. Different representations of the theoretical progress curve obtained from the numerical solution of the ODE system comprising Eqs. 1.7-1.9 using $S_0/K_m = 1$, $E_0/K_m = 0.01$ and $K_S/K_m = 1$. (A) Product concentration $[P]$ represented over time t in a linear plot. Red tangent lines represent the period of time τ_n that would be required to complete the reaction if the instant reaction rate was maintained. For long reaction times this period of time tends to the value of the characteristic time constant τ_∞. The slope of the initial tangent corresponds to the value of initial reaction rate v_0. (B) Log-linear plot of the concentration of product still to be formed $(S_0 - P)$ as a function of time. The slope of final tangent (red dashed line) corresponds to the negative reciprocal of the characteristic time constant.

The asymptotic limit of $(S_0 - P)/v$ for late reaction phases is here defined as the characteristic time constant τ_∞ and corresponds to the reciprocal of the "integration constant" shown in the original MM paper to be independent of the initial substrate concentration [1]. Later interpretation of steady-state results identified the integration constant as the specificity constant k_2/K_m (or, more generically, k_{cat}/K_m) multiplied by the enzyme concentration [12], while its reciprocal corresponds to the period of time τ needed to completely exhaust the existing substrate if the initial reaction rate is maintained and the enzyme is operating under first-order conditions [13]. Despite the similarities between the latter definition and our own definition of τ_∞, the following differences should be noted beforehand: the time constant τ is defined in relation to the initial reaction rates under steady-state conditions, whereas τ_∞ is concerned with the late reaction phases under whatever experimental conditions. From the definition of the pivotal variable for long reaction times given in the Supporting Information of the PEA paper [9], the following relationship exists between τ_∞ and τ (Eq. 1.10):

$$\tau_\infty = \frac{\tau}{2}\left(1 + e_0 + \sqrt{(1 + e_0)^2 - 4\left(1 - \frac{K_S}{K_m}\right)e_0}\right)$$

(1.10)

The representation of this function in Figure 1.2A takes into account the alternative definition of $1/\tau$ as $k_2 e_0$ to show that the shortest characteristic time corresponds to $1/k_2$ and is obtained for enzyme concentrations above the Michaelis constant. This compromise between finishing reaction rates and enzyme concentration motivated us to propose an efficiency index ϕ balancing kinetic performance over the enzyme expenditure:

$$\phi = \frac{1/\tau_\infty}{E_0}$$

(1.11)

Defined in this way, enzyme efficiency is exempted from the practical limitations of the specificity constant, whose application to compare the catalytic efficiency of different enzymes in the catalysis of the same substrate has been discouraged [14]. In fact, by attending to the final phases of the enzymatic reaction, the definition of ϕ is free from the ambiguities caused by the role of the substrate concentration on the initial reaction rates [14]. On the other hand, the fact illustrated in Figure 1.2B that the maximum value of efficiency ϕ_{max} corresponds to the value of k_2/K_m (or, more generically to k_{cat}/K_m) might be extremely convenient so as to recover published k_{cat}/K_m data from any misgivings while comparing the efficiency of different enzymes. Finally, and as addressed more in detail in Discussion, the efficiency index can be straightforwardly estimated from a single enzymatic assay using Eq. 1.11 and the values of τ_∞ determined as described in Figure 1.1B.

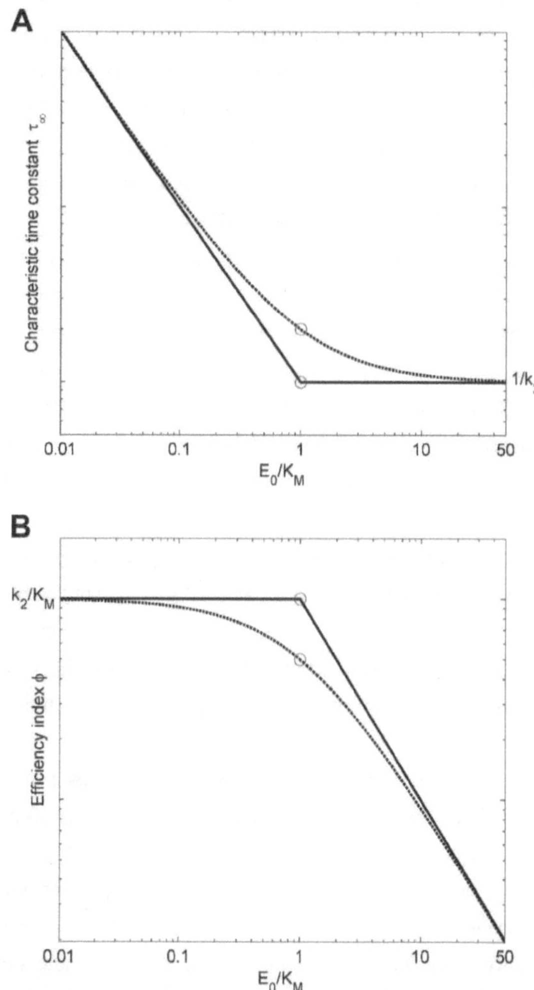

Figure 1.2. The characteristic time constant and enzyme efficiency. Log-log plots depicting the influence of the K_m-normalized enzyme concentration on the (A) characteristic time constant τ_∞ the (B) efficiency index ϕ for limiting values of the scaled dissociation constant $K_S/K_m = 0$ (solid lines) and $K_S/K_m = 1$ (dashed lines). The blue round markers show the point where the largest difference between both curves is observed. (A) The smallest value for the characteristic time is obtained for $E_0 > K_m$. (B) The maximal efficiency index ϕ_{max} is obtained for $E_0 < K_m$.

In the original MM paper, the now-called Michaelis constant K_m was defined as the protein-ligand dissociation constant [1], which for enzyme-substrate complexes is now commonly represented by K_S. Comparing their mathematical formulations given in Section 1.2 shows that the catalytic step (rate constant k_2) must be much slower than the unbinding step (rate constant k_{-1}) for K_m to be equivalent to K_S [15]. In the PEA paper, K_S is referred to as a non-MM constant, which, together with K_m and V, completes the portrayal of the 3-parameter mechanism proposed by Briggs and Haldane [9]. Figure 1.3 shows two sets of theoretical curves simulated for enzyme concentrations much lower than K_m (Figure 1.3A) and equal to K_m (Figure 1.3B) to illustrate the peculiar role of K_S in both situations. Figure 1.3A partly explains the absence of K_S from steady-state kinetic analysis, seeing that the enzyme-substrate affinity has a weak effect on the progress curves, which is only visible for product conversions below 5%, and considering substrate concentrations S_0 close to E_0. This does not mean that K_S is equivalent to K_m, only that the effect of K_S is masked under conditions of great substrate excess. In the other extreme, experimental conditions for which the enzyme concentration is of the same order of magnitude of K_m (and $S_0 \leq E_0$) are expected to clearly reveal the effect of K_S during initial and late phases of the progress curves [9]; for this reason, and because of the biological interest, this is considered a "critical region of conditions" that is potentially representative of an intracellular environment [16-18]. Figure 1.3B shows that asymptotically high affinities between enzyme and substrate ($K_S/K_m = 0$) should produce characteristic product accumulation curves with sigmoidal (rather than hyperbolic/linear) onsets.

Since low K_S/K_m ratios mean much faster product formation rates than enzyme-substrate dissociation rates, it might be technically difficult to access the earlier phases of such kinetic curves and discern their shape, especially when high enzyme concentrations are involved. The PEA alternative to estimate the value of K_S/K_m is through the characteristic time constant τ_∞, which, as described in the previous subsection, can be straightforwardly obtained from a single enzymatic assay. Given that the characteristic time constant is independent of the initial substrate concentration, values of S_0 as high as the solubility limit can be adopted in order to extend the duration of the catalytic reactions over technically accessible time periods.

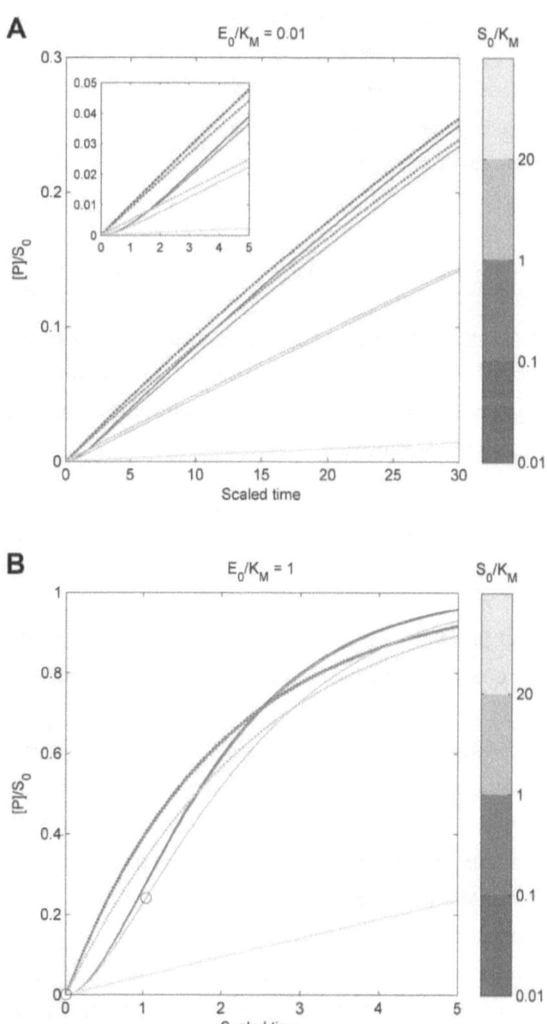

Figure 1.3. Major differences between theoretical progress curves calculated for limiting values of the dissociation constant $K_S/K_m = 0$ (solid lines) and $K_S/K_m = 1$ (dashed lines). Progress curves represented as the linear plots of the normalized product concentration $[P]/S_0$ over the scaled time $\theta = k_2 t$. The system of ODE comprising Eqs.1.7-1.9 was solved using the set of S_0/K_m values indicated in the log-scaled colorbars for (A) $E_0/K_m = 0.01$ and (B) $E_0/K_m = 1$. (B) The blue round markers on the curves obtained for $S_0/K_m = 1$ indicate the stationary moment for which the maximum reaction velocity is reached.

As previously represented in Figure 1.2A, the influence of E_0 on τ_∞ is not significantly affected by the value of the K_S/K_m ratio, unless enzyme concentrations close to K_m are considered. This window of conditions is, therefore, recommended to estimate the dissociation constant from experimentally determined characteristic time constants. The K_S/K_m value follows directly from Eq. 1.10 rewritten as Eq. 1.12:

$$\frac{K_S}{K_m} = 1 + (k_{cat}\tau_\infty)^2 \frac{E_0}{K_m} - (k_{cat}\tau_\infty)\left(1 + \frac{E_0}{K_m}\right) \tag{1.12}$$

which requires previous estimations of the MM parameters using, for example, the PEA model Eq. A1.2 in Appendix or the MM equation (Eq. 1.6 for steady-state conditions only). In Discussion we anticipate some of the practical and fundamental consequences arising from the accurate knowledge of the parameter K_S.

1.5. Discussion

The present work is the first follow-up of the PEA model, which, as the acronym incidentally suggests, is envisaged to *seed* several other future applications in modern enzymology. Specifically, we took the opportunity presented by the 7th ESCEC Symposium to expand on the meaning and practical significance of the characteristic time constant τ_∞ and of the equilibrium dissociation constant K_S. The relevance of these parameters was already intuited in the 1915 paper of MM, seeing that $1/\tau_\infty$ and K_S correspond, in the limit cases, to the original "integration constant" and to the Michaelis constant, respectively. More than enlarging the steady-state scope, our approach motivates a renewed interpretation of the fundamental meaning of MM and non-MM kinetic constants. For example, enzyme efficiency defined in relation to τ_∞ is not affected by the concentration of substrate and, therefore, it is free from the ambiguities associated to the specificity constant defined as the k_{cat}/K_m ratio extracted from initial velocity experiments [14]. A direct indicator of the enzyme's kinetic performance, the value of $1/\tau_\infty$ is also an apparent first-order rate constant that increases with the concentration of enzyme until the upper limit of k_2 is attained for $E_0 > K_m$ (Figure 1.2A). Consequently, enzyme efficiency is here presented as the kinetic performance balanced over the total enzyme expenditure $\phi = 1/(\tau_\infty E_0)$. Figure 1.2B showed that the efficiency index reaches a maximal value of k_2/K_m that is nearly invariant for enzyme concentrations below K_m. This value of ϕ_{max}, which operationally corresponds to k_{cat}/K_m, can be used to compare the catalytic effectiveness of different enzymes for technological applications or for enzyme evolution studies. As the differences summarized in Table 1.1 intend to illustrate, the numerical equivalence between ϕ_{max} and the specificity constant is circumstantial and does not imply a common underlying principle. Different fundamental definitions (#4 and #5 in Table 1.1) stipulate different methodological procedures for the determination of the two indicators (#1 to #3 in Table 1.1) which, nevertheless, should produce the same numerical results, provided that the Briggs-Haldane mechanism holds true. Reaction schemes involving product inhibition may originate different values of k_{cat}/K_m if estimated as ϕ_{max} or as a specificity constant (#6 in Table 1.1). Although product inhibition is not contemplated by the PEA model, the common usage of apparent rate constants (such as k_{cat}) as an approximation to true rate constants (such as k_2) might also be extended to the efficiency index, whose apparent value may help to characterize quantitatively the deviations from Briggs-Haldane kinetics.

Table 1.1. Different interpretations of k_{cat}/K_m in the light of the MM model (as a specificity constant) and in the light of the PEA model (as the maximal enzyme efficiency ϕ_{max}). Differences 1 to 3 concern parameter estimation methodologies; differences 4 and 5 concern kinetic and operational meanings, respectively; difference 6 concerns reaction schemes other than Briggs-Haldane.

#	Specificity Constant	ϕ_{max}
1	Estimated based on *initial* reaction rates v_0	Estimated based on the characteristic time constant τ_∞ during *late* reaction phases
2	*Limited* to **steady-state** experimental conditions	Estimations of ϕ are *not limited* to any experimental condition; ϕ_{max} is reached for $E_0 < K_m$
3	Substrate concentration *influences* the v_0 - based enzyme's efficiency [14]	Substrate concentration *does not* influence τ_∞-based enzyme's efficiency
4	Corresponds to an apparent second-order rate constant	Corresponds to an apparent first-order rate constant expressed per units of enzyme concentration
5	Sets the *lower* limit for enzyme-substrate association rate constant [19]	Sets the *upper* limit of the ratio enzyme performance/enzyme expenditure
6	It is *not* affected by product inhibition	It may be affected by product inhibition

Another MM parameter subject to a renewed PEA perspective is the Michaelis constant itself. Appointed as less important than parameters k_{cat} and k_{cat}/K_m [12], the value of K_m is frequently defined as the concentration of substrate producing $v_0 = 0.5 V_{max}$; on the other hand, the formulation $K_m = (k_{-1} + k_2)/k_1$ indicates that the Michaelis constant is an overall/apparent dissociation constant of all enzyme-bound species [19]. The latter definition is directly concerned with the enzyme-substrate affinity, which can be characterized accurately using true dissociation constants (K_S) determined as described in the previous subsection.

The PEA model additionally shows that the first definition of K_m (as the substrate concentration yielding half-maximal rates) loses its validity outside the region of steady-state conditions [9]. For example, for $E_0 > S_0$ the initial reaction rate v_0 becomes linear dependent on the substrate concentration in the cases of very low enzyme-substrate affinity ($K_S/K_m \sim 1$) – see Eq. A1.2b in Appendix. Instead, Figure 1.2 confers to parameter K_m the biophysical significance of a threshold enzyme concentration. According to Figure 1.2A, K_m is the smallest enzyme concentration required to achieve the shortest completion time, i.e., required to conclude the enzymatic reaction at the fastest rates. Perhaps more useful for *in vivo* and *in vitro* kinetic analysis, Figure 1.2B presents K_m as the maximum enzyme concentration that can be kept without losing catalytic efficiency – after this limit, increasing enzyme expenditure no longer accelerates the concluding reaction phases. Curiously enough, enzyme concentrations close to the value of K_m are also the most favorable to experimentally investigate the effect of the enzyme-substrate dissociation constant on the characteristic time constant – Eq. 1.10. According to this new angle of approach, enzyme efficiency can be regulated by dynamically controlling the enzyme's abundance in the cell. Concentration levels close to the reference value of K_m are important for the enzyme to be critically sensible to the structural affinity of different metabolites. By systematically adopting steady-state conditions, it is conceivable that *in vitro* enzymatic assays have been missing kinetic aspects of metabolic homeostasis that are important [9], for example, in molecular systems biology [20] and in drug discovery [21-23].

The enzyme-substrate affinity is important to define which catalysis occurs preferentially in a cellular environment crowded with multiple enzymes and substrates that possibly act as competitors towards each other. Therefore, the explanation for the apparent disregard of the dissociation constant K_S comparatively to k_{cat} or K_m resides in the lack of straightforward methods to estimate this non-MM constant. Existing methods for the determination of all individual rate constants require specific techniques designed to measure transient-state kinetics, the interpretation of which is not exempted from simplifying hypothesis such as the RSA during the pre-steady-state phases [8,19] or the linearization of the reaction mechanism for time-relaxation analysis [5]. These limitations are not present in the PEA method for the determination of K_S using the characteristic time constant and Eq. 1.12. By facilitating the characterization of enzyme specificity, we also expect to contribute to the understanding of enzyme evolution and enzyme promiscuity, upon which the design of novel biological functions is based [24]. A quantitative description of the enzyme response to alternative substrates is now possible using true dissociation constants as an alternative to entropic predictions based on the k_{cat}/K_m ratio [25].

1.5.1. A Single Assay to Estimate Enzyme Activity, Efficiency and Affinity (EA)²

Estimating the MM parameters requires different enzymatic reactions to be carried out adopting substrate concentrations S_0 above and below K_m and in great excess over the enzyme $(S_0 \gg E_0)$. Although the usage of a single progress curve to determine K_m and V is theoretically possible, this procedure is discouraged in practice in view of the undefined time span over which the SSA is valid [26]. The insights provided by the PEA model let us envisage a new method to determine the classic parameters from a single enzymatic reaction and in an unbiased manner. In addition, the information thus obtained can be used to analyze a second progress curve in order to estimate the non-MM parameter K_S. Because this method characterizes enzyme activity, efficiency and affinity we call it the (EA)² assay. In principle, the (EA)² assay involves the following steps:

1. Measure the progress curve of the enzymatic reaction under typical steady-state conditions $(S_0 \gg E_0)$
2. Determine the initial reaction rate v_0 as indicated in Figure 1.1A
3. Determine the characteristic time constant τ_∞ as indicated in Figure 1.1B. Assume that
$$\tau_\infty = \tau$$
4. Estimate V from Eq. 1.6 (Eq. MM) rewritten as $V = v_0 S_0/(S_0 - v_0\tau)$
5. Estimate $K_m = V\tau$
6. The condition $\tau_\infty = \tau$ in Step 3 is only valid for $E_0 \ll K_m$ (Figure 1.2A). Check if $E_0 < 0.1K_M$
    a. If not, restart with a more diluted enzyme solution
7. Estimate enzyme's activity as V/E_0 (equivalent to k_{cat})
8. Estimate the maximal enzyme's efficiency $\phi_{max} = 1/(\tau_\infty E_0)$ (corresponding to k_{cat}/K_m for the conditions of step 6)
9. Measure a new progress curve adopting $E_0 = K_m$ and determine a new value of τ_∞
10. Estimate the dissociation constant K_S characterizing the enzyme-substrate affinity. Use the value of τ_∞ estimated in step 9 and Eq. 1.12 rewritten as $K_S/K_m = (1 - k_{cat}\tau_\infty)^2$

Notably, this method does not require knowing an accurate value of the substrate concentration S_0, provided that this value is assuredly much higher than the product $v_0\tau_\infty$ so as to obtain $V = v_0$ in Step 4. The MM parameters can alternatively be determined using the PEA model Eq. A1.2 in the Appendix or the MM equation (Eq. 1.6 for steady-state conditions only). When the enzyme molarity is not accurately known, the (EA)² assay might also be useful to estimate the lower limit of the catalytic power taking into consideration that E_0 estimates such as absorbance readings at 280 nm are in excess, thus yielding lower limits of enzyme activity V/E_0 and of enzyme efficiency $1/(\tau_\infty E_0)$. In another instance, if only the amount of impure powdered enzyme is known, enzyme efficiency can be expressed in units of $s^{-1}(mg/l)^{-1}$ as an alternative to $s^{-1}M^{-1}$, similarly to what happens with the catalytic activity expressed as the amount of enzyme converting the substrate into product at a given rate (1 mol/s or 1 μmol/min for katal or international unit IU, respectively). It may occur that the $(EA)^2$ assay fails to produce useful data because of either too slow or too fast enzymatic reactions; in the first case, sample conditions may not be maintained with time (e.g. protein degradation leading to enzyme-activity loss); in the latter case, the reaction may finish before any valid measurement is performed – especially under the $E_0 = K_m$ conditions of step 9. The solution to these problems involves decreasing or increasing the substrate concentration values within the operational limits in order to prolong or shorten the reaction span to convenient limits. Obtaining enzyme samples as concentrated as the K_m-order of magnitude might also not be possible in practice. In those cases, the estimation of the K_S/K_m ratio is still possible using the initial phases of the progress curves measured using dilute enzyme solutions [9]. During the application of the PEA model and, in particular, of the $(EA)^2$ assay, the Briggs-Haldane mechanism is implicitly assumed to be valid. As previously discussed in Table 1.1, deviations from this mechanism can be identified by comparing the estimations of MM parameters obtained from initial and late phases of the enzymatic reactions using, in one case, Eq. A1.2 in Appendix, and in the other, the characteristic time constant τ_∞. We intend to keep developing the ideas organized in this paper by applying them on the characterization of enzymatic systems with biological and industrial interest.

1.6. <u>Conclusions</u>

Firstly published in the same year of the classic MM paper, Marcel Proust's novel *À la Recherche du Temps Perdu*, In Search of Lost Time (1913-1927), gives the motif for the title of the present contribution, in which we try to recuperate the fundamental meanings of the characteristic time constant τ_∞ and of the equilibrium dissociation constant K_S. This exercise is based on the recently published PEA model that provides, after a long wait, the closed-form solution of the Briggs-Haldane kinetic mechanism [9]. Although the Briggs-Haldane mechanism is the minimal reaction scheme needed to explain enzyme catalysis, it remained very incompletely described by the existing analytical solutions. The pivotal variable of the PEA model measured for late reaction phases gives a practical estimate of the characteristic time constant τ_∞, which in turn is helpful to clarify the concepts of enzyme efficiency and selectivity.

The maximal enzyme efficiency ϕ_{max} corresponds to the value of $1/(\tau_\infty E_0)$ measured for concentrations of enzyme below K_m (Figure 1.2B). Parameter ϕ_{max} is expected to help in recovering the wealth of published k_{cat}/K_m data from the criticism it has been voted as an efficiency standard: although both parameters are, in most cases, numerically equivalent, ϕ_{max} is free from the conceptual limitations of k_{cat}/K_m (Table 1.1). The PEA framework also provides a renewed perspective of the somewhat obscure Michaelis constant K_m as a threshold enzyme concentration above which the catalytic efficiency starts to decrease. The practical definition of K_m as the substrate concentration yielding half-maximal rates should be adopted carefully as it loses accuracy under non-steady-state conditions. The true dissociation constant K_S can now be straightforwardly determined from a single progress curve without requiring specific experimental arrangements or model simplifications. Besides completing the Briggs-Haldane portrayal of the catalytic cycle, this parameter objectively characterizes the affinity of the enzyme to different substrates, thus contributing to the study of enzyme evolution and promiscuity. Summarizing our conclusions, a practical method for determining enzyme activity, efficiency and affinity from single progress curves is proposed, in which model parameters are rapidly estimated even if the concentrations of substrate and enzyme are not accurately known.

1.7. Appendix

The following equations comprise the overall and stationary formulations of the PEA model as described by Pinto *et al.* in 2015 [9] (see Introduction, Section II.I). The overall analytical solution corresponds to Eq. A1.1a, where scaled variables are used, namely $\tau = K_m/V$, $e_0 = E_0/K_m$, $s_0 = S_0/K_m$, $\theta = t/(e_0\tau)$ and $\beta = 1 - K_S/K_m$.

$$\frac{S_0 - P}{v} = \frac{\tau}{2}\left(1 + e_0 + \tilde{s} + \frac{\tilde{\lambda}}{tanh\left(\frac{\tilde{\lambda}\theta}{2\beta}\right)}\right) \tag{A1.1a}$$

The corresponding daughter variables $\tilde{s}$, $\tilde{\lambda}$, s^* and θ^* are given by Eqs. A1.1b-A1.1e. The value of λ^* in Eq. A1.1e corresponds to the value of $\tilde{\lambda}$ calculated by Eq. A1.1c for $\tilde{s} = s^*$. The superscript asterisk is indicative of stationary conditions, occurring after the initial fast transient period of $[ES]$ build-up has taken place.

$$\tilde{s} = \omega\left(s^* exp\left(s^* - e_0(\theta - \theta^*)\right)\right) \tag{A1.1b}$$

$$\tilde{\lambda} = \sqrt{(1 + e_0 + \tilde{s})^2 - 4\beta e_0} \tag{A1.1c}$$

$$s^* = \frac{1}{2}\left(s_0 - 1 - e_0 + \sqrt{(s_0 + e_0 + 1)^2 - 4e_0 s_0}\right) \tag{A1.1d}$$

$$\theta^* = \frac{2\beta \, arctanh\left(\frac{\lambda^*}{1 + e_0 + s^*}\right)}{\lambda^*} \tag{A1.1e}$$

The choice of the stationary instant t^* is in order to simplify the usage of the PEA model given that the stationary pivotal variable $(S_0 - P^*)/v^*$ is independent of K_S. The stationary version of the PEA model can easily be used to estimate MM parameters through the application of linear regressions [9]:

$$\frac{S_0 - P^*}{v^*} = \begin{cases} \dfrac{K_m + S_0}{V}, & S_0 > E_0 \\[2mm] \dfrac{K_m + E_0}{V}, & S_0 < E_0 \end{cases} \tag{A1.2a}$$

It should be noted that in the case of maximal dissociation constant ($K_S/K_M = 1$), the previous equation is reduced to the MM equation for $S_0 > E_0$, and to the simplified Bajzer and Strehler equation [27] for $S_0 < E_0$:

$$v_0 = \begin{cases} \dfrac{VS_0}{K_m + S_0}, & S_0 > E_0 \\[2mm] \dfrac{VS_0}{K_m + E_0}, & S_0 < E_0 \end{cases} \tag{A1.2b}$$

References

1. Michaelis, L. and Menten, M. (1913), *Die Kinetik der Invertinwirkung.* Biochemische Zeitschrift. 49:333–369.
2. Brown, A.J. (1902), *XXXVI.—Enzyme action.* Journal of the Chemical Society, Transactions. 81:373-388.
3. Henri, V. (1902), *Théorie générale de l'action des quelques diastases.* Comptes rendus hebdomadaires des séances de l'Académie des sciences. 135:916-919.
4. Henri, V., (1903), *Lois Générales de l'action des Diastases.* Librairie Scientifique A. Hermann (Paris).
5. Cornish-Bowden, A., (2012), *Fundamentals of Enzyme Kinetics*, 4th ed. Wiley-Blackwell (Weinheim, Germany), pp.25-45.
6. Briggs, G.E. and Haldane, J.B.S. (1925), *A note on the kinetics of enzyme action.* Biochemical Journal. 19:338-339.
7. Berberan-Santos, M.N. (2010), *A General Treatment of Henri-Michaelis-Menten Enzyme Kinetics: Exact Series Solution and Approximate Analytical Solutions.* MATCH Communications in Mathematical and in Computer Chemistry. 63:283-318.
8. Hanson, S.M. and Schnell, S. (2008), *Reactant Stationary Approximation in Enzyme Kinetics.* The Journal of Physical Chemistry A. 112:8654-8658.
9. Pinto, M.F., *et al.* (2015), *Enzyme kinetics: the whole picture reveals hidden meanings.* FEBS Journal. 282(12):2309-2316.
10. Segel, L.A. (1988), *On the validity of the steady state assumption of enzyme kinetics.* Bulletin of Mathematical Biology. 50(6):579-593.
11. Dormand, J.R. and Prince, P.J. (1980), *A family of embedded Runge-Kutta formulae".* Journal of Computational and Applied Mathematics. 6:19-26.
12. Johnson, K.A. and Goody, R.S. (2011), *The Original Michaelis Constant: Translation of the 1913 Michaelis-Menten Paper.* Biochemistry. 50:8264-8269.
13. Cornish-Bowden, A. (1987), *The Time Dimension in Steady-state Kinetics: A Simplified Representation of Control Coefficients.* Biochemical Education. 15(3):144-146.
14. Eisenthal, R., Danson, M.J., and Hough, D.W. (2007), *Catalytic efficiency and k_{cat}/K_M: a useful comparator?* TRENDS in Biotechnology. 25(6):247-249.
15. Baici, A., (2015), *Kinetics of Enzyme-Modifier Interactions*, 1st ed. Springer, pp.39.
16. Schnell, S. and Maini, P.K. (2000), *Enzyme kinetics at high enzyme concentration.* Bulletin of mathematical biology. 62(3):483-499.
17. Tzafriri, A.R. (2003), *Michaelis-Menten kinetics at high enzyme concentrations.* Bulletin of mathematical biology. 65(6):1111-1129.

18.   Bersani, A.M. and Dell'Acqua, G. (2011), *Asymptotic expansions in enzyme reactions with high enzyme concentrations.* Mathematical Methods in the Applied Sciences. 34(16):1954-1960.

19.   Fersht, A., (1999), *Structure and Mechanism in Protein Science*, 2nd ed. W.H. Freeman and Company (New York).

20.   Finn, N. and Kemp, M., (2014), Systems Biology Approaches to Enzyme Kinetics: Analyzing Network Models of Drug Metabolism. Methods in Molecular Biology. Vol. 1113. Humana Press, Chp. 15.

21.   Acker, M.G. and Auld, D.S. (2014), *Considerations for the design and reporting of enzyme assays in high-throughput screening applications.* Perspectives in Science. 1(1–6):56-73.

22.   Yang, J., Copeland, R.A., and Lai, Z. (2009), *Defining balanced conditions for inhibitor screening assays that target bisubstrate enzymes.* Journal of Biomolecular Screening. 14(2):111-20.

23.   Sols, A. and Marco, R. (1970), *Concentrations of metabolites and binding sites. Implications in metabolic regulation.* Current Topics in Cellular Regulation. 2:227–273.

24.   Pandya, C., *et al.* (2014), *Enzyme Promiscuity: Engine of Evolutionary Innovation.* The Journal of Biological Chemistry. 289(44):30229-30236.

25.   Nath, A. and Atkins, W.M. (2008), *A quantitative index of substrate promiscuity.* Biochemistry. 47(1):157-166.

26.   Duggleby, R.G. (2001), *Quantitative analysis of the time courses of enzyme-catalyzed reactions.* Methods. 24(2):168-174.

27.   Bajzer, Z. and Strehler, E.E. (2012), *About and beyond the Henri-Michaelis-Menten rate equation for single-substrate enzyme kinetics.* Biochemical and Biophysical Research Communications. 417(3):982-985.

Chapter 2.

A simple linearization method unveils hidden enzymatic assay interferences

The content of the present chapter constitutes the second peer-reviewed article published in the context of the present book (Article II, see Annex 2):

Pinto, M.F., Ripoll-Rozada, J., Ramos, H., Watson, E.E., Franck, C., Payne, R.J., Saraiva, L., Pereira, P.J.B., Pastore, A., Rocha, F., Martins, P.M. (2019), A simple linearization method unveils hidden enzymatic assay interferences. *Biophysical Chemistry*. 252:106193. doi:10.1016/j.bpc.2019.106193

2.2. <u>Introduction</u>

Typically, more than one-third of the discrete drug targets in the portfolio of pharmaceutical companies consists of enzymes [1], with phosphate-transferring enzymes, or kinases, being the largest category of potentially novel drug targets [2]. Drug screening is usually based on enzymatic assays that aim at identifying compounds that inhibit, enhance or modulate enzyme activity. However, the output of these assays strongly depends on the experimental conditions and on several different parameters that are often difficult to master completely. In high-throughput screening (HTS) of enzyme modulators, primary assays employing light-based detection methods are escorted by orthogonal assays using different output reporters in order to identify false positives and fluorescence/ luminescence artifacts [3,4]. Other possible interferences can be specific of a given system, such as the occurrence of enzyme inactivation and competitive product inhibition, or unspecific, as in the cases of random experimental errors and of changes in experimental parameters during the reaction (Figure 2.1). This uncertainty dramatically calls for new and more sensitive approaches to allow fast and reliable detection of these interferences. While no kinetic method is currently available to detect generic interferences in enzymatic assays, in the specific case of enzyme inactivation interferences their occurrence can be detected by the Selwyn test applied to progress curves measured at different enzyme concentrations (E_0) and constant substrate concentration (S_0) [5]. Yet, besides requiring the realization of additional experiments, the Selwyn test provides no quantitative information of inactivation rates [6] and might not detect incomplete enzyme inactivation. In the case of nonspecific aggregation interferences in HTS assays, counter-screens of β-lactamase inhibition in the presence and absence of detergent are performed to check for the presence of promiscuous inhibitors [7].

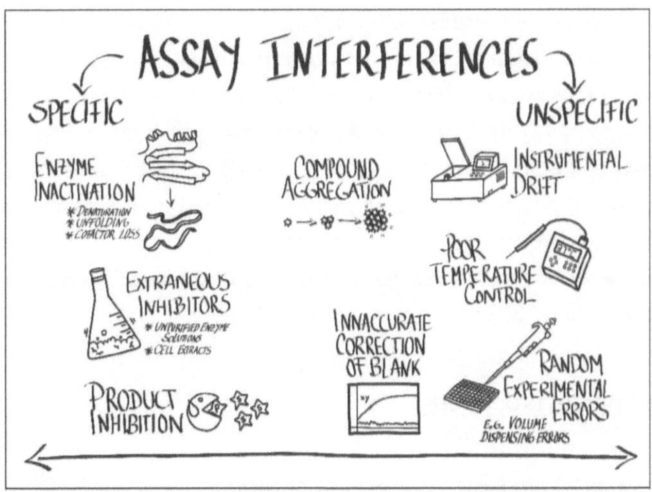

Figure 2.1. Specific and unspecific interferences on enzymatic assays. Spurious effects that are too small to be readily observable can produce important errors of interpretation of kinetic results. Specific of a given system, enzyme inactivation can be prevented by the addition of protein stabilizers or by increasing the concentration of enzyme [8]. Extraneous inhibitors present in unpurified enzyme solutions or in cell extracts systematically affect the quality of kinetic measurements [9]. Product inhibition can usually be ignored in initial-rate measurements but is highly misleading in time-course studies [10]. Compound aggregation can cause enzyme sequestration on the surface of the aggregate particles and is one of the main reasons for promiscuous enzyme inhibition [3,11]. Instrumental drift, poor temperature control, inaccurate correction of the sample blank, and random experimental errors in e.g. volume dispensing operations are typical examples of unspecific interferences [12,13]. Adequate buffering of the reaction mixture is important to prevent changes of enzyme activity provoked by drifts in pH and ionic strength [14].

In the present contribution, we propose a touchstone criterion for the detection of assay interferences based on the graphical representation of reaction coordinates in a linearized scale. We applied our method to enzymatic reactions catalyzed by procaspase-3, caspase-3 (EC 3.4.22.56) and α-thrombin (EC 3.4.21.5). Caspases are a family of cysteine-dependent aspartate-specific peptidases (MEROPS family C14; [15]) synthesized as zymogens and converted into their more active forms upon proteolytic cleavage [16]. Both caspase-3 and its precursor procaspase-3 undergo progressive inactivation during *in vitro* enzymatic assays. Progress curves of procaspase-3- and caspase-3-catalyzed reactions are analyzed to identify enzyme inactivation and characterize its relative importance. Alpha-thrombin is a (chymo)trypsin-like serine peptidase (MEROPS family S01; [15]) and a main effector in the coagulation cascade. Similar to caspase-3, its zymogen (prothrombin) is cleaved to generate the active form of the enzyme. Thrombin generation is tightly regulated to allow blood clot formation after an injury [17]. A variety of thrombin-targeting inhibitors is produced by blood-feeding organisms [18-21]. The outcome of the new test in the presence of enzyme inhibition is demonstrated for α-thrombin-catalyzed reactions inhibited by a synthetic variant of an anticoagulant produced by *Dermacentor andersoni* [22,23]. Along with the inactivation and inhibition studies, we discuss the detection of unspecific interferences arising from changes in the reaction conditions.

2.3. <u>Experimental procedures</u>

2.3.1. Procaspase-3 production in yeast cell extracts

Procaspase-3 was produced as previously described [24,25]. Briefly, cultures of *Saccharomyces cerevisiae* transformed with the expression vector pGALL-(*LEU2*) encoding human procaspase-3 were diluted to 0.05 optical density at 600 nm (OD$_{600}$) in 2% (w/v) galactose selective medium and grown at 30 °C under continuous shaking until an OD$_{600}$ range of 0.35-0.40. Cells were collected by centrifugation and frozen at −80 °C. For protein extraction, cell pellets were thawed, treated with *Arthrobacter luteus* lyticase (Sigma-Aldrich), and the cells were lysed using CelLytic™ Y Cell Lysis Reagent (Sigma-Aldrich) in the presence of 1 mM phenylmethylsulfonyl fluoride, 1 mM Dithiothreitol (DTT), 1 mM ethylenediaminetetraacetic acid (EDTA), 0.01% (v/v) Triton X-100. Total protein concentration of the extracts was determined using the Pierce™ Coomassie Protein Assay Kit (ThermoFisher Scientific).

2.3.2. Enzymatic assays for procaspase-3 and caspase-3

The activity of recombinant human procaspase-3 (STRENDA ID 1XV0MK) and of recombinant human purified caspase-3 (STRENDA ID M9FKPY) was followed by monitoring the conversion of the fluorogenic substrate Acetyl-Asp-Glu-Val-Asp-7-amido-4-methylcoumarin (Ac-DEVD-AMC) to 7-amino-4-methylcoumarin (AMC) at 37 °C. Procaspase-3 (0.123 mg/mL protein extract) and caspase-3 (1.0 U, Enzo Life Sciences) were assayed in 96-well microplates

(Nunc™ MicroWell™ 96-Well, Thermo Scientific™, ThermoFisher Scientific) using 100 µL of 20 mM HEPES pH 7.4 (20 °C), 100 mM NaCl, 10% (w/v) sucrose, 0.1% (w/v) CHAPS, 10 mM DTT, 1 mM EDTA per well [26]. A range of substrate concentrations of 3.125-300 µM Ac-DEVD-AMC for procaspase-3, and 3.125-50 µM Ac-DEVD-AMC for caspase-3 were tested. The reactions were started by addition of protein, and fluorescence was monitored at 460 nm (390 nm excitation) using a HIDEX CHAMELEON V plate reader (Turku, Finland). To avoid evaporation, the reaction mixture in each well was overlaid with liquid paraffin (100 µL). All solutions were equilibrated to 37 °C before use. Assays were performed in triplicate or quadruplicate for procaspase-3, and in duplicate for purified caspase-3. The calibration curve was built using solutions with known concentration of the free fluorescent product AMC (Sigma-Aldrich).

2.3.3. Enzymatic assay for α-thrombin

The enzymatic activity of human α-thrombin (0.15 nM, Haematologic Technologies) was assessed by following its amidolytic activity toward the chromogenic substrate Tos-Gly-Pro-Arg-p-nitroanilide (25-400 µM, Chromozym TH, Roche) at 37 °C in the presence of 0.40 nM of a synthetic variant of an anticoagulant produced by *D. andersoni* [23] - STRENDA ID OUYDF2. The assays were performed at least in duplicate in 200 µL of 50 mM Tris pH 8.0 (20 °C), 50 mM NaCl, 1 mg/ml BSA. The reactions were initiated by the addition of α-thrombin and followed at 405 nm using a Synergy2 multi-mode microplate reader (BioTek) [27].

2.4. The linearization method

Deviations from the normal progress of enzyme-catalyzed reactions should, in principle, alter the build-up profile of product concentration (P) vs. time (t) from the theoretical curve expected by the integrated form of the Michaelis-Menten (MM) equation [28-30]:

$$P = Vt + K_m \ln\left(1 - \frac{P}{S_0}\right) \tag{2.1}$$

where S_0 is the initial substrate concentration and K_m and V are the Michaelis constant and the limiting rate, respectively. In practice, however, Eq. 2.1 is not used to detect assay interferences since no evident changes in the shape of the progress curves are induced by enzyme inactivation, product inhibition, and quasi-equilibrium mechanisms of competitive inhibition, uncompetitive inhibition, etc. [31]. Eq. 2.1, which in its closed-form version is also known as the Schnell-Mendoza equation [28], produces poorer estimates of kinetic parameters than the classic MM equation fitted to initial reaction rate measurements because any external interference may severely accumulate over the full time-course [29,30]. In addition, the application of Eq. 2.1 is limited to a range of conditions more restricted than that of $E_0 \ll S_0 + K_m$ required to validate the steady-state assumption [32,33]: Eq. 2.1 fails to account for the fast transient phase that precedes the steady-state phase according to the closed-form solution of the Briggs-Haldane reaction mechanism (Figure 2.2a). Additionally, the effect of E_0 can be masked by apparent values of the Michaelis constant ($K_m^{app} \approx K_m + E_0$), especially for $E_0 \geq K_m$ [34].

We propose a new linearization method (LM) for the detection of assay interferences based on the following modified version of the integrated MM equation:

$$\frac{\Delta P}{\Delta t} = V^{app} - \left[-K_m^{app} \ln\left(1 - \frac{\Delta P}{\Delta P_\infty}\right) \middle/ \Delta t \right] \tag{2.2}$$

The main differences of this formalism relatively to Eq. 2.1 are: (a) the use of apparent kinetic constants K_m^{app} and V^{app}, (b) the use of partial time intervals ($\Delta t = t - t_i$) and of the corresponding increment of product concentration ($\Delta P = P - P_i$), (c) the initial condition (subscript i) is now any point of the reaction subsequent to the pre-steady-state phase ($t_i > t^*$), (d) the final concentration of product is given by the measured value (P_∞) and not by the expected value (S_0) [30], and (e) Eq. 2.2 is presented in a linearized form of the Walker and Schmidt type [35] (Figure 2.2b). Features (a) to (c) are meant to expand the validity of time-course analysis to the same range of conditions of the steady-state assumption ($E_0 \ll S_0 + K_m$) [34,36] (see Introduction, Section I.I, and Chapter 1, Section 1.5). Feature (d) takes into account possible discrepancies between P_∞ and S_0 values resulting, for example, from complete enzyme inactivation or inaccurate pipetting. Feature (e) is implemented because the Walker and Schmidt linearization [35] is highly sensitive to fluctuations in P readouts [37], whereas linearity is an easily implementable judgment criterion.

The linear variation of $\Delta P/\Delta t$ with $-\ln(1 - \Delta P/\Delta P_\infty)/\Delta t$ expected by Eq. 2.2 is a necessary but not sufficient condition to reject assay interferences. To pass this test, the straight lines obtained at different substrate concentrations should also superimpose each other (Figure 2.2b). Although parameter estimation is not the primary goal here, the agreement between the apparent constants and the values of K_m and V obtained by the standard initial-rate method further confirms that the assay is unbiased. Finally, since the LM equation applies to single active site, single substrate and irreversible steady-state reactions of the Briggs-Haldane type

[38], this method might also be used to reveal the presence of a more complex enzymatic mechanism from the one assumed.

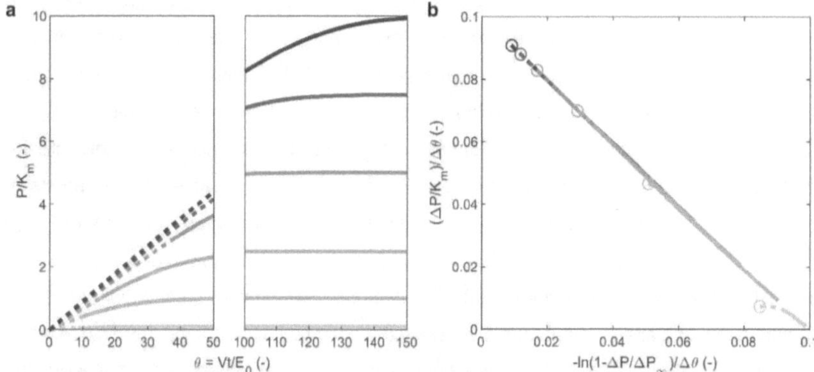

Figure 2.2. Theoretical progress curves expected for unbiased Briggs-Haldane reaction mechanisms. (a) Evolution of product concentration over time represented in dimensionless units of product concentration ($p = P/K_m$) and time ($\theta = Vt/E_0$) for different substrate concentrations. Different colors correspond to values of $s_0 = S_0/K_m$ of (from top to bottom) 10, 7.5, 5, 2.5, 1, 0.1 (additional simulation parameters given in Appendix A Section 2.8.1, and Table A2.1, Appendix A Section 2.8.5). The broken x axis is used to emphasize the initial periods of constant velocity (dashed lines). (b) The same curves are represented in linearized coordinates according to Eq. 2.2; round markers indicate the steady-state instant t^*. The absence of assay interferences is evidenced by negatively-sloped, superimposing straight lines.

2.5. Results and discussion

2.5.1. Procaspase-3 inactivation - preliminary analysis

The exponential reaction curves of Ac-DEVD-AMC cleavage by procaspase-3 (Figure 2.3a) are not suggestive of any evident loss of enzyme activity over time. The lack of well-defined slopes from which the initial rates (v_0) can be accurately measured might only indicate that the substrate concentrations are still too low to achieve the saturating MM conditions [39,40]. In fact, the plateau corresponding to V in the MM representation (Figure 2.3b) is barely noticeable in the studied range of substrate concentrations. The obtained value of $K_m = 217 \pm 59$ µM is 62-fold higher than that previously reported using the uncleavable mutant procaspase-3(D$_3$A), which has three processing sites removed, and the substrate Ac-DEVD-AFC, which has a different fluorescent reporter (AFC) than Ac-DEVD-AMC [41]. Differences between the observed and the literature values of K_m could therefore be ascribed to the distinct nature of each enzymatic assay. Yet, the calibration curve represented in Figure 2.3b (inset) provides important clues as to the possible existence of experimental artifacts. This is a conditional calibration curve since the final fluorescence value (RFU_∞) is assumed to result from a complete catalytic reaction in which the final product concentration P_∞ is equivalent to S_0. To check the validity of this hypothesis, reference fluorophore solutions were used to calibrate the equipment according to the standard protocol (Figure 2.4a); for the same concentrations of fluorophore and substrate, the fluorescence intensity of the calibration solutions (RFU_{cal}) clearly surpasses the RFU_∞ signal, thus suggesting partial conversion of substrate into products during the reactions (Figure 2.3a). This finding invalidates the preliminary calibration curve and the kinetic analysis in Figure 2.3b since the condition of complete chemical reaction is not observed. Whether or not the reactions were really unfinished and what were the mechanisms thereby involved cannot be ascertained by the calibration-curve test alone.

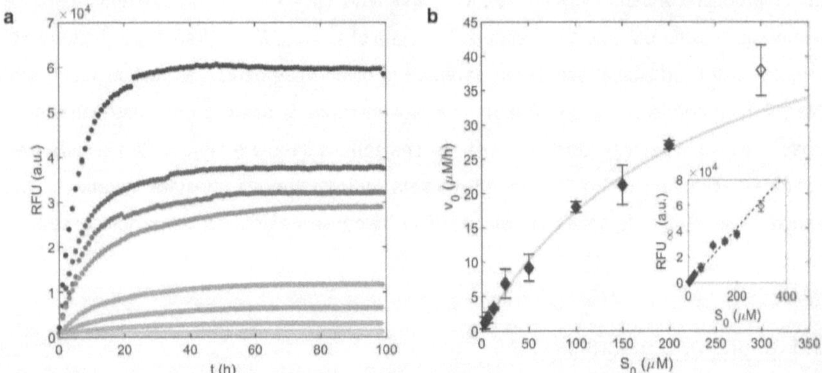

Figure 2.3. Loss of procaspase-3 activity is not self-evident in a first analysis of kinetic results. (a) Fluorescence (RFU) increase during the cleavage of Ac-DEVD-AMC by procaspase-3 for S_0 values of (from top to bottom) 300, 200, 150, 100, 50, 25, 12.5, 6.25, and 3.125 µM. (b) Plot of the initial reaction rates (v_0) as a function of substrate concentration. Symbols and error bars: means and standard deviations of v_0 values calculated using the initial slopes obtained in (a) and a preliminary calibration curve relating the end-point fluorescence (RFU_∞) and S_0 (inset). Solid line: fit of the MM equation to selected experimental data (closed symbols). Since the results are affected by severe enzyme inactivation, the fitted values of $K_m = 217 \pm 59$ µM and $V = 55 \pm 9$ µM/h ($R^2 = 0.9919$) are merely indicative. Inset: linear fit (dashed line) to selected (closed symbols) RFU_∞ vs. S_0 data ($R^2 = 0.9532$).

Besides procaspase-3 inactivation, other interfering factors, such as fluorescence quenching phenomena, could have caused the observed differences between RFU_{cal} and RFU_{∞}. The occurrence of progressive loss of enzyme activity was further confirmed by the Selwyn test (Figure 2.4b) and by the direct measurement of procaspase-3 activity for different periods of incubation in the reaction environment (Figure 2.4c). Relatively to these methods, the calibration-curve test (Figure 2.4a) has the advantage of requiring no other experiments than those already performed while estimating MM kinetic parameters. Furthermore, if enzyme inactivation is admitted as a first-order decay process, the following relationship can be used to quantitatively estimate the product of the decay rate constant λ by the time constant $\tau = K_m/V$ (Appendix A Section 2.8.2):

$$P_{\infty} = S_0 - K_m \omega \left[\frac{S_0}{K_m} \exp \left(\frac{S_0}{K_m} - \frac{1}{\tau \lambda} \right) \right] \tag{2.3}$$

For a given substrate concentration, the extent of the reaction is ultimately determined by the product $\tau \lambda$ relating the rates of inactivation and of unimpaired reaction. It follows from the inverse dependence of $\tau \lambda$ on E_0 (via τ and V) that the effects of inactivation can be counterbalanced by increasing the concentration of enzyme. Similar improvements can be achieved by decreasing the value of λ through the use of protein stabilizers. The fitted value of $\tau \lambda = 1.4 \pm 0.3$ is clearly in the region above ~0.1 for which complete enzyme inactivation can be attained before the conversion of the total available substrate (view Tables A2.2 and A2.3, Appendix A Section 2.8.5). Contrary to the estimation obtained by initial-rate analysis, the fitted value of $K_m = 161 \pm 77$ µM is obtained taking into account the effect of enzyme inactivation. The quantitative information provided by the calibration-curve test is an advantage over the Selwyn test, whose underlying principle also requires additional progress curves to be measured with various enzyme concentrations and constant S_0 values [5]. In the case described in Figure 2.4b, the non-superimposed Selwyn plots of P against $E_0 t$ confirm the likely occurrence of procaspase-3 inactivation, as previously suggested by simple inspection of the calibration curves (Figure 2.4a). The last evidence supporting the verdict of both tests is obtained by directly measuring the enzymatic activity at the end of different periods of incubation in the reaction environment (Figure 2.4c). Procaspase-3 activity is confirmed to rapidly decrease with time according to the exponential decay trend expected for first-order processes. The reciprocal of the decay rate constant ($1/\lambda = 8.3$ h) can be interpreted as the period of time required for the catalytic activity to drop to ~37% of its initial value.

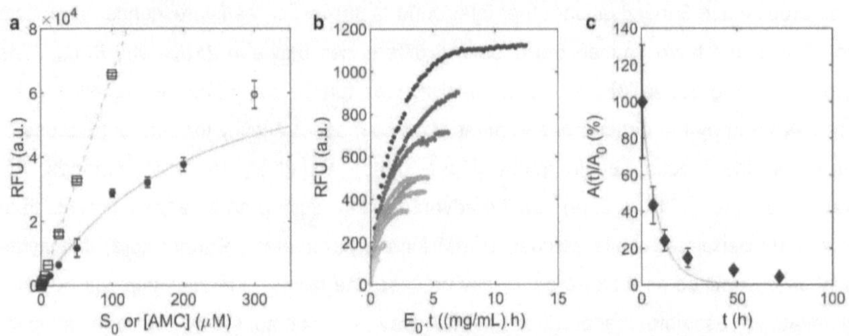

Figure 2.4. Procaspase-3 inactivation identified by different methods. (a) The calibration-curve test is proposed based on the differences between *expected* (RFU_{cal}, squares) and *obtained* (RFU_∞, circles) end-point signals. Symbols and error bars represent means and standard deviations. The values of RFU_{cal} are measured using solutions of known fluorophore concentrations ([AMC]). The test passes if the values of RFU_{cal} and RFU_∞ superimpose. Solid line: Eq. 2.3 is fitted to the experimental data (fitting results: $K_m = 161 \pm 77$ μM and $\tau\lambda = 1.4 \pm 0.3$ h, $R^2 = 0.9851$). The obtained end-point signals are the same used to build the preliminary calibration curve in the inset of Figure 2.3b. Dashed line: linear fit representing the true calibration curve ($R^2 = 1$). (b) The classic Selwyn test also suggests time-dependent loss of procaspase-3 activity since progress curves measured for (from top to bottom) 0.17, 0.13, 0.09, 0.06, 0.07, 0.06, 0.04, 0.02, 0.01 mg/mL procaspase-3 and constant S_0 (3.125 μM) are not superimposable when represented in a modified $E_0 t$ timescale [5]. (c) Symbols and error bars: means and standard deviations of the normalized enzymatic activity $A(t)/A_0$ after different periods of incubation. Solid line: numerical fit to an exponential decay function (fitted result: $\lambda = 0.12 \pm 0.02$ h⁻¹, $R^2 = 0.9655$).

2.5.2. Loss of procaspase-3 activity identified by the linearization method

Complete catalytic reactions, promoted for example by high E_0 values or by low S_0 values, can pass the calibration-curve and the Selwyn tests even in presence of significant enzyme inactivation. This misjudgment may affect the overall quality of reported enzymology data, particularly when the initial reaction rate phase is itself difficult to define, as in conditions of $S_0 < K_m$ [39,42]. Enhanced limits of detection to this and other interferences can be achieved by application of the new LM test brought forward by Eq. 2.2. In the commonest case of $E_0 \ll K_m$, the intervals Δt and ΔP can be considered right from the beginning of the measurements because the steady-state condition starts to be valid after the first milliseconds of the reaction [34,43]. The non-conformity of the linearized curves of procaspase-3 (Figure 2.5) with the ideal behavior previously described in Figure 2.2b is evident: the linearized progress curves obtained for different values of S_0 show positive slopes (rather than negative slopes) and do not superimpose.

The calibration-curve test, the Selwyn test and the LM test are all capable of detecting the progressive inactivation of procaspase-3. For practical uses, the new methods here proposed are more easily applicable than having to prearrange and perform additional Selwyn test experiments. However, the question remains open as to which of the three methods is more sensitive to slight losses of enzymatic activity. In order to study milder decay processes than that observed for procaspase-3 in cell extracts, values of $\tau\lambda < 1.4$ were used in the numeric simulations described in detail in the Appendix A (Tables A2.2-A2.4, Section 2.8.5). For a reference value of $\tau\lambda = 0.1$ and assuming S_0/K_m ratios between ~1 and ~5, both the calibration-curve test (Figure 2.6a) and the Selwyn test (Figure 2.6b) fail to reveal S_0-dependent effects caused by inactivation. Conversely, these effects are visibly amplified in Figure 2.6c where the $(\Delta P/K_m)/\Delta\theta$ vs. $-\ln(1 - \Delta P/\Delta P_\infty)/\Delta\theta$ curves are at once non-linear and non-superimposing. Therefore, our experimental and numeric results confirm that the LM test is a practical, yet stringent, alternative to detect enzyme inactivation.

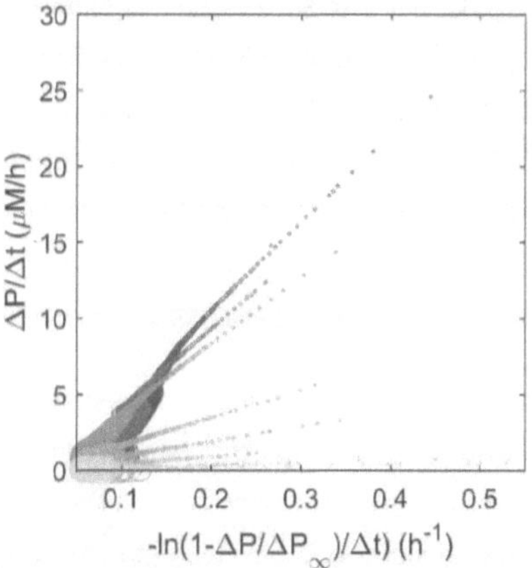

Figure 2.5. Using the LM test to detect procaspase-3 inactivation. The progress curves in Figure 2.3a measured for S_0 values of 200, 150, 100, 50, 25, 12.5, 6.25, and 3.125 µM are now represented in the linearized $\Delta P/\Delta t$ vs. $-\ln(1 - \Delta P/\Delta P_\infty)/\Delta t$ scale (color-coded as in Figure 2.3a). Symbol size increases with the time-course of the reaction. Non-superimposing, positively-sloped straight lines clearly indicate the occurrence of assay interferences, which, in the present case, are associated with procaspase-3 inactivation (Figure 2.4).

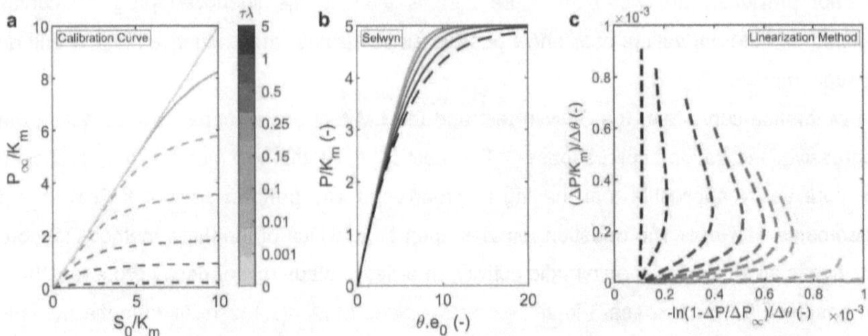

Figure 2.6. The LM test is highly sensitive to enzyme inactivation. Detection limits of the (a) calibration-curve test, (b) Selwyn test and (c) the LM test according to the theoretical progress curves simulated assuming a first-order decay of enzyme activity (simulation parameters listed in Tables A2.2-A2.4, Appendix A Section 2.8.5). Dashed and solid lines: cases of successful and failed detection, respectively. (a) The calibration-curve test fails to detect inactivation for $\tau\lambda$ values (color bar) below 0.1, as the measurable and expected values of final product concentration (P_∞ and S_0) start to be undistinguishable. (b) Theoretical progress curves simulated for varying e_0 values for a range of $\tau\lambda$ values between 10^{-3} and 0.1 (from lighter to darker shades of gray), and $S_0/K_m = 5$. The Selwyn test fails to detect inactivation for $\tau\lambda < 0.1$, as the Selwyn curves are not easily distinguishable. (c) Theoretical LM curves simulated for a reference value of $\tau\lambda = 0.1$ and a range of S_0/K_m values between (from lighter to darker shades of gray) 0.1 and 10. This test detects inactivation under conditions of $\tau\lambda = 0.1$ and $S_0/K_m < 5$ for which the calibration-curve test and the Selwyn test have poor sensitivity.

2.5.3. Unspecific interferences detected in the caspase-3 assay

Faster catalytic reactions are expectable for purified caspase-3 relatively to procaspase-3, which, besides being less active, is present in low concentration in cell extracts. Consequently, the inactivation issues considered for the proenzyme are less important for the purified enzyme – note that the duration of the enzymatic reactions decreases from > 1 day (Figure 2.3a) to < 1 h (Figure 2.7a). New challenges for the accurate determination of kinetic parameters are, however, posed by the shorter reaction timescales. This is illustrated in Figure 2.7a, where the phase of constant velocity is not clearly defined during a stabilization period of ~10 min. Short periods of normalization of reaction conditions are hardly avoidable even when, as in the present case, the component solutions are pre-equilibrated to the reaction temperature, or when miniaturized high-throughput devices are employed [44]. Small temperature variations markedly influence enzymatic reaction rates and in a S_0-dependent manner [45]. After confirming that the caspase-3 assay is not affected by significant enzyme inactivation (Figure 2.7b) the reaction phases that succeed the first ~ 10 min interval can be analyzed in detail. Because the initial slopes (and the corresponding values of v_0) are probably affected by drifts in the reaction properties, instantaneous rates (v_i) obtained upon condition stabilization may be used in v vs S_0 plots as an approximation to the real value of v_0 (Figure 2.7c). For rapidly progressing reactions, this procedure raises the doubt of whether the instantaneous substrate concentration is too depleted relative to the initial value (S_0) [46]. Also, it is not granted that the properties of the reaction mixture are completely stabilized during the period of time over which v_i is determined.

A better perception of the main experimental outliers can be obtained by representing the progress curves in the modified scale proposed by the LM test (Figure 2.8a). For each S_0 condition, the initial measurements stand out as evidently separated from the negatively-sloped trend exhibited by most of the subsequent data points. This suggests that the values of v_0 determined from Figure 2.7a (open circles) and the resulting kinetic analysis in Figure 2.7c (dotted line) are affected by assay interferences. The fact that no straight line common to all S_0 conditions is clearly defined by the late data points does not have a particular meaning because early errors can propagate throughout the $\Delta P / \Delta t$ vs. $-\ln(1 - \Delta P / \Delta P_\infty)/\Delta t$ curve. Moreover, the final amplification of the instrumental noise is expectable in result of the use of the logarithm in the horizontal axis. Setting the new initial condition to $t_i = 10$ min not only eliminates the initial outliers but also improves the quality of the linearized plots (Figure 2.8b). The tendency of the different experimental curves to superimpose in a single straight line further confirms that the activity of caspase-3 remains practically unchanged during the time-course of the reactions. Overall, the first (Figure 2.8a) and second (Figure 2.8b) LM representations validate the enzymatic assay for purified caspase-3, although the initial ~ 10 min stabilization period should not be considered for analysis. In the determination of instantaneous reaction rates, the used $P(t)$ data (closed circles in Figure 2.7a) already integrate the negatively-sloped trends in Figure 2.8a. The kinetic laws based on v_i measurements (Figure 2.7c, solid line) should, however, take into account the depletion of substrate until the moment when the rate is determined [34]. This correction to the MM plots involves replacing the initial substrate concentration S_0 by the instantaneous substrate concentration S_i, here estimated using median concentration values for the time interval used for v_i determination. The MM parameters fitted to the v_i vs. S_i data (Figure 2.8c; $K_m = 21.5$ µM and $V = 109$ µM/h) are considered valid and free from major assay interferences. In accordance to what is expected for steady-state conditions, the apparent kinetic constants K_m^{app} and V^{app} of the LM equation can be approximated by the true parameters K_m and V.

Illustrating this, the experimental LM curves (symbols in Figure 2.8b) are well described by the theoretical LM curve computed for $K_m^{app} = K_m$ and $V^{app} = V$ (dashed line in Figure 2.8b).

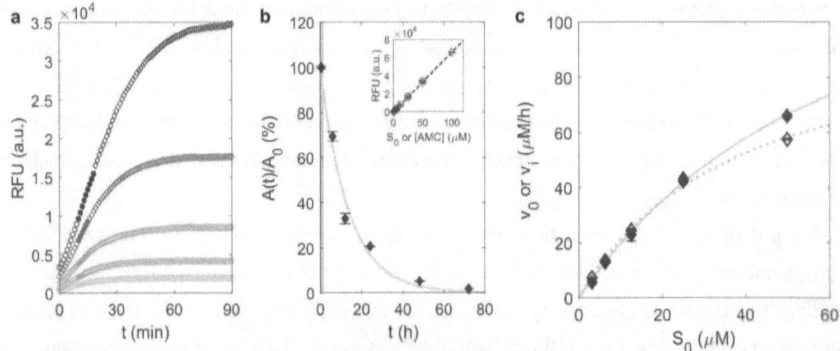

Figure 2.7. Assay interferences other than enzyme inactivation may affect the initial-rate measurements in the reactions catalyzed by purified caspase-3. (a) Fluorescence increase during the hydrolysis of Ac-DEVD-AMC by 1.0 U of caspase-3 for S_0 values of (from top to bottom) 50, 25, 12.5, 6.125, and 3.125 µM. Circles: data selected for the determination of initial (open symbols) and instantaneous (closed symbols) slopes. (b) Symbols and error bars: means and standard deviations of direct measurements of caspase-3 activity after different periods of incubation in the reaction environment. No significant inactivation occurs within the full reaction timescale (grey area) considered in (a). Line: Numerical fit to an exponential decay function ($\lambda = 0.074 \pm 0.009$ h^{-1}, $R^2 = 0.9829$). Inset: the calibration-curve test confirms the absence of significant enzyme inactivation: the obtained end-point signals (circles) overlie the calibration curve (dashed line) built with fluorescence measurements of standard AMC solutions (squares). (c) Plot of the initial (v_0) and instantaneous (v_i) reaction rates as a function of the initial substrate concentration (S_0). The experimental values of v_0 (open symbols) and v_i (closed symbols) are calculated using initial and instantaneous slopes, respectively, as represented in (a). Lines: fit of the MM-like equation to the experimental data. Since both v_0 and v_i are imperfect estimations of the initial rate value (see text for details) the fitted values of $K_m = 38.7 \pm 6.4$ µM and $V = 103 \pm 10$ µM/h (dotted line, open symbols, $R^2 = 0.9971$) and $K_m = 71.9 \pm 13.0$ µM and $V = 162 \pm 20$ µM/h (solid line, closed symbols, $R^2 = 0.9982$) are merely indicative.

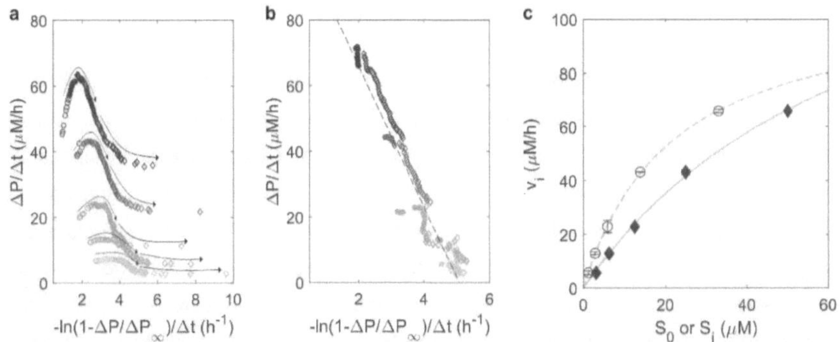

Figure 2.8. Time-wise variations in solution properties are detected by the LM test. (a) Symbols: progress curves of Ac-DEVD-AMC catalysis by 1.0 U caspase-3 represented in the linearized scale using $t_i = 0$ min (color-coded as in Figure 2.7a). Arrows: visual reference indicating the reaction time-course. The initial experimental outliers (open circles) show up detached from the negatively-sloped trends. The final scattering of the data results from the amplification of random errors and instrumental noise. (b) Symbols: linearized progress curves obtained after discarding the initial outlier points ($t_i = 10$ min). For clarity, only data corresponding to 95% reaction completion are represented. Dashed line: representation of Eq. 2.2 after replacing K_m^{app} and V^{app} by the values of K_m and V determined by independent methods in (c). (c) Symbols and error bars: means and standard deviations of the instantaneous reaction rates (v_i) represented as a function of S_0 (closed symbols) and S_i (open symbols). Lines: numerical adjustment of the MM-like equation. The v_i vs. S_0 data and numerical fit (closed symbols and solid line) are the same as in Figure 2.7c (shown here as visual reference). Dashed line: numerical fit of the v_i vs. S_i data; the fitted results ($K_m = 21.5 \pm 0.9$ μM and $V = 109 \pm 2$ μM/h, $R^2 = 0.9998$) are not affected by major assay interferences as they successfully describe the experimental LM curves in (b).

2.5.4. Inhibition of α-thrombin

The presence of unaccounted enzyme modifiers in the assay solution is another possible interference associated to the use of crude enzyme preparations and cell extracts. The detection of enzyme modulation effects by the LM test is here demonstrated for the inhibition of the amydolytic activity of human α-thrombin by a synthetic variant of an anticoagulant produced by *D. andersoni*. Since the inhibitory effect of subnanomolar concentrations of the enzyme-modifier is known beforehand (Figure A2.1, Appendix A Section 2.8.4), the LM test is applied to identify the fingerprints left by enzyme modifiers and to illustrate how cautiously initial-rate measurements should be used during the characterization of inhibition mechanisms.

If (i) quasi-equilibrium conditions are rapidly attained and (ii) the concentration of inhibitor is significantly higher than the concentration of enzyme, progress curves measured in the presence of competitive, uncompetitive or mixed inhibition are still numerically described by Eq. 2.2, with K_m^{app} and V^{app} being affected by the concentration of inhibitor(s). Consequently, the LM analysis may fail to detect linear inhibition effects provoked by solution contaminants. On the positive side, quasi-equilibrium linear inhibition is a particular case of the general modifier mechanism [47] (see Chapter 4, Sections 4.2.2 and 4.2.3), whose rate equations frequently contain squared concentration terms recognizable as deviations from the LM equation. As such, it is conceivable that the new method can also be used for the preliminary detection of variants of the general modifier mechanism, and not only for assay validation purposes. The complexity of this subject greatly increases as other LM-detectable mechanisms of product inhibition, slow-onset inhibition, substrate competition, allosterism, etc., are considered. Presently, we apply the LM test as a quality-control test to α-thrombin-catalyzed reactions inhibited by a synthetic variant of an anticoagulant produced by *D. andersoni*, and we leave the fundamental characterization of the inhibition mechanism to future research (see Appendix B for further analysis and discussion). This model system is useful to illustrate how the presence in subnanomolar amounts (0.40 nM) of a given compound might be revealed by characteristic kinetic signatures left in LM curves. The sigmoid-shaped onset of the α-thrombin progress curves (Figure 2.9a) is admissible under conditions of $S_0 \leq E_0$ that do not apply here [34] (see Chapter 1, Section 1.4.2). Once again, the ill-defined initial rates can admittedly result from the gradual stabilization of the experimental conditions and not necessarily from the presence of enzyme modifiers. Yet, unlike what was observed for capase-3, the first LM representation (Figure 2.9b) suggests that stabilization periods much longer than 10 min are required for the emergence of negatively-sloped LM curves. Even admitting a stabilization period of 30 min in the second LM representation (Figure 2.9c), no superimposable trend is clearly defined by the individual curves obtained at the different S_0 conditions. This means that, in the case of α-thrombin, imperfect temperature control during the initial reaction phases cannot be used to explain the inconsistent LM curves obtained afterwards. Taken together, the long initial periods evidencing positive LM slopes and the persisting lack of a well-defined trend common to the different S_0 conditions indicate possible deviations from the basic Briggs-Haldane mechanism. After dismissing the hypothesis of enzyme inactivation (which can be done using the calibration curve test), the presence of high-affinity enzyme modifiers is a strong possibility to be considered even when their action is through a direct inhibition mechanism. In fact, rate equations containing squared concentration terms are not exclusive of hyperbolic modifiers but are also expected for tight-binding linear inhibitors occurring at concentrations in the order of magnitude of E_0 or lower [48-50].

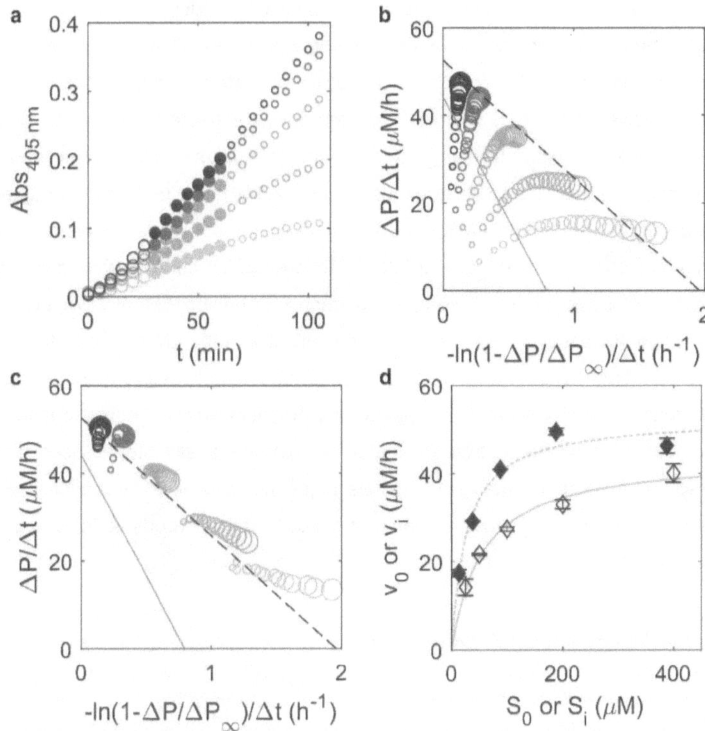

Figure 2.9. The LM test detects timewise effects induced by enzyme modifiers. Absorbance (Abs_{405nm}) increase during the catalysis of Tos-Gly-Pro-Arg-pNA by 0.15 nM α-thrombin in the presence of 0.40 nM inhibitor for substrate concentration of (from top to bottom) 400, 200, 100, 50 and 25 μM. Large symbols: data selected for the determination of initial (large open symbols) and instantaneous (closed symbols) slopes. (b) The same progress curves are represented in the linearized scale using $t_i = 0$ min. (c) Symbols: LM progress curves obtained after discarding the initial points ($t_i = 30$ min). (b,c) Lines: representation of Eq. 2.2 after replacing K_m^{app} and V^{app} by the values of K_m and V determined in (d) using initial (solid line) and instantaneous (dashed line) measurables. Symbol size increases with the time-course of the reaction. (d) Symbols and error bars: means and standard deviations of the initial reaction rates (v_0) and instantaneous reaction rates (v_i) represented as a function of S_0 (open symbols) and S_i (closed symbols), respectively. Lines: using the MM-like equation to fit v_0 vs. S_0 data (solid line, $K_m = 55.8 \pm 9.3$ μM and $V = 44.1 \pm 2.3$ μM/h, $R^2 = 0.9929$) and v_i vs. S_i data (dashed line, $K_m = 26.9 \pm 7.2$ μM and $V = 52.6 \pm 3.4$ μM/h, $R^2 = 0.9839$). The LM test is not passed because the individual trends in (c) fail to converge into a single overall straight line.

For this reason, unsuspected contaminants that are also tight-binding modifiers will have their effects uncovered by representing the measured progress curves in LM coordinates. Although the saturation plots obtained using initial (index 0) or instantaneous (index i) variables correspond to typical MM curves (Figure 2.9d), the fitted parameters are, in both cases, of ambiguous physical meaning. The v_0 vs. S_0 analysis (solid lines in Figure 2.9b and Figure 2.9c) clearly does not pass the LM test, as the theoretical LM curve (solid line in Figure 2.9c) fails to intercept the experimental LM curves. When instantaneous measurables are analyzed (dashed lines in Figure 2.9c and Figure 2.9d), the theoretical LM curve is able to intercept, at least in part, the experimental results obtained for each S_0 condition (Figure 2.9c); even so, the α-thrombin enzymatic assay is considered non-compliant with the LM prerequisites because the individual LM time-course trends (indicated by symbol size increase in Figure 2.9c) are divergent from the overall straight line suggested by the theoretical LM curve (dashed line in Figure 2.9c). This didactic example serves to demonstrate that assay interferences cannot be diagnosed solely based on the quality of numerical adjustments to the MM equation. In contrast, kinetic effects caused by very small amounts of either linear or hyperbolic inhibitors can be detected by the LM test.

2.5.5. The LM test as a routine quality check

The LM test is suitable for routine use in enzymatic assay validation and to decide which optimization steps should be taken in order to improve reproducibility and accuracy. As summarized in Figure 2.10, its application is based on the LM representations of reaction progress curves obtained for a fixed value of E_0 and varying S_0. First, the coordinates $\Delta P/\Delta t$ and $-\ln(1 - \Delta P/\Delta P_\infty)/\Delta t$ are computed using the beginning of the measurements ($t_i = 0$ min) as initial condition for identifying the initial period of stabilization of reaction conditions. If the obtained LM curves are negatively-sloped straight lines and tend to superimpose since the beginning of the reaction, the test is passed. If evident deviations from the ideal trend are found to occur, new coordinates must be computed using as initial reference any instant subsequent to the period of stabilization ($t_i > 0$ min). If the LM test is still not passed, the presence of enzyme inactivation can be tested using the calibration-curve test. Some assay interferences can be minimized by increasing the concentration and/or purity of the enzyme; the presence of unspecific assay interferences must also be checked here. This process can be repeated iteratively until the enzymatic assay is fully optimized. It should also be noted that repeated failure to optimize the enzymatic assay using this method can indicate the presence of a kinetic mechanism that deviates from that of Briggs-Haldane. Further kinetic studies using complementing methodologies should be considered in such case. Next, we present additional practical guidelines to be considered during the systematization of the new method.

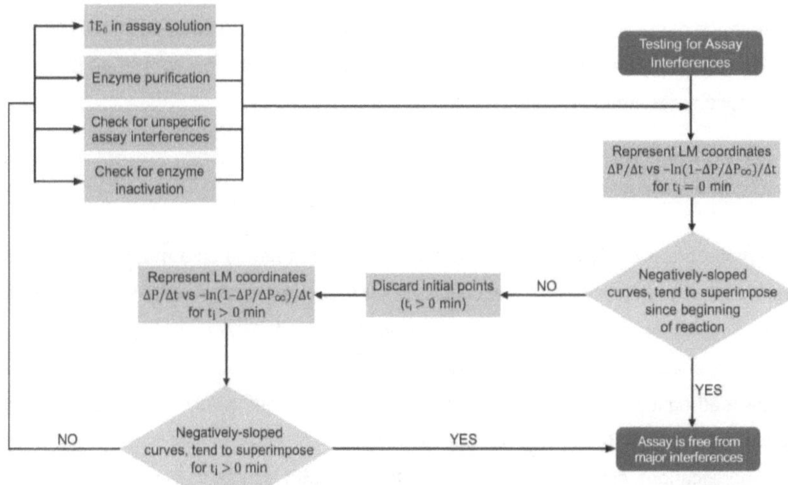

Figure 2.10. Flowchart depicting the main steps of the LM test for the identification of assay interferences.

More complete information is provided by the LM test when full progress curves are measured. Note, for example, that the $\Delta P/\Delta t$ coordinate does not change during the whole initial period of constant velocity. If, in a limit case, all points are collected within constant velocity timeframes, no individual trends will be defined for each substrate concentration and the result of the test will be solely dictated by the linearity of the overall trend. Time-dependent changes in enzymatic activity can thus be missed when only a part of the reactions is analyzed. Letting the reaction proceed until its end also allows identifying the differences between P_∞ and S_0 values upon which the calibration-curve test is based. Although less sensitive than the LM test, the calibration-curve test is specific for enzyme inactivation and, for this reason, is indicated for the preliminary assessments of this type of interference. Nonetheless, as shown for the case of α-thrombin, reaction progress curves that reach a final plateau indicating reaction completion are not mandatory for the application of the LM test.

Small errors in reagent handling give rise to differences between the values of P_∞ and S_0 that are only evident at the end of the reaction. Such unspecific interferences are detected by the LM test even if the final product concentrations are not known, and S_0 values have to be used instead of P_∞ to compute the $-\ln(1 - \Delta P/\Delta P_\infty)/\Delta t$ coordinate. In the numerical example given in the Appendix A (Figure A2.2, Section 2.8.4), a random error of 5% in the value of S_0 generates, since the beginning of the reaction, a clear deviation of the biased LM curve relatively to the overall trend defined by the other LM curves. As the reaction progresses in time, the differences become more pronounced and even the initial linearity is lost. A consequence of the Walker and Schmidt-type linearization [30,35] adopted by the LM equation, this high responsiveness to random experimental errors is helpful for controlling the validity of each reaction rate measurement.

2.6. Conclusions

A new method to identify assay interferences is proposed based on a modified version of the integrated MM equation. To pass the so called "LM test", progress curves measured at different substrate concentrations and represented in linearized $\Delta P/\Delta t$ vs. $-\ln(1 - \Delta P/\Delta P_\infty)/\Delta t$ coordinates should superimpose in a single, negatively-sloped straight line. The proposal of this new method follows from the recently obtained closed-form solution of the Briggs-Haldane reaction mechanism [34,36] (see Introduction, Section II, and Chapter 1, Section 1.5). Some of the modifications now introduced to the integrated MM equation allow time-course kinetic analysis to be carried out in the range of conditions commonly adopted for steady-state analysis ($E_0 \ll S_0 + K_m$). The illustrative examples of enzymatic reactions catalyzed by procaspase-3, caspase-3 and α-thrombin highlight different aspects that can stealthily influence the quality of enzymology data. Initial rate measurements during the catalysis of Ac-DEVD-AMC by procaspase-3 in yeast cell extracts are strongly affected by progressive enzyme inactivation, promptly detected by the LM test independently of whether such suspicion exists *a priori* or not. The Selwyn test, a reference method to identify enzyme inactivation [5,51], is less sensitive than the LM test and requires additional experiments whenever enzyme inactivation is somehow suspected to occur. The catalysis of Ac-DEVD-AMC by purified caspase-3 was used to demonstrate that non-conformities in the first LM representation are a possible indication that the experimental conditions were not yet stabilized at the beginning of the measurements. In fact, the LM representation obtained after discarding the initial stabilization period validated the caspase-3 assay, which was then used to determine unbiased MM parameters based on instantaneous values of substrate concentration and reaction rate. Finally, the inhibition of α-thrombin by subnanomolar concentrations of a synthetic anticoagulant showed the interest of the LM test in amplifying subtle enzyme modifier effects. This example illustrated that high-affinity contaminants may affect enzyme kinetics in a hard to detect way unless LM plots are represented and individual LM curves are compared with the overall trend.

Because stringent criteria are adopted, nonstandard catalytic reactions (characterized by multiple active sites, multiple substrates, product inhibition, hyperbolic inhibition, tight-binding inhibition, enzyme inactivation, etc.), or that are influenced by instrumental noise and poor automatic control will probably not pass the LM test. This fine quality control is crucial for the success of quantitative kinetic analysis since important mechanistic nuances "may play a subordinated role with respect to even modest mistakes in reagent handling, (...) instrumental noise and others" [52, p. 108]. In line with current initiatives to improve enzymology data reporting [53,54], this new method is expected to contribute in improving the reproducibility of kinetic data, thus increasing the impact of fundamental and applied research in fields such as enzyme engineering, systems biology and drug discovery.

2.7. <u>Appendix A</u>

2.7.1. Irreversible enzyme inactivation - numeric solutions

The Briggs-Haldane reaction scheme comprises the binding of substrate (S) to free enzyme (E) to give rise to the enzyme-substrate complex (ES), from which the catalytic product (P) is formed and released, thereby regenerating the free enzyme. Irreversible enzyme inactivation can be accounted for in this scheme by considering that free and bound enzyme decay irreversibly into the inactive forms (E^* and ES^*). For simplicity we will assume that both processes are well described by the same first-order rate constant (λ):

$$E + S \underset{k_{-1}}{\overset{k_1}{\rightleftharpoons}} ES \overset{k_2}{\rightarrow} E + P$$

$$E^* \qquad\qquad ES^*$$

This reaction scheme is mathematically described by the following system of first-order differential equations:

$$\frac{d[S]}{dt} = k_{-1}[ES] - k_1[E][S] \tag{A2.1a}$$

$$\frac{d[ES]}{dt} = k_1[E][S] - (k_{-1} + k_2 + \lambda)[ES] \tag{A2.1b}$$

$$\frac{d[E]}{dt} = (k_{-1} + k_2)[ES] - k_1[E][S] - \lambda[E] \tag{A2.1c}$$

$$\frac{d[P]}{dt} = k_2[ES] \tag{A2.1d}$$

subject to the initial conditions $([S], [ES], [E], [P]) = (S_0, 0, E_0, 0)$ and to the mass conservation laws $E_0 = [E] + [ES] + [E^*] + [ES^*]$ and $S_0 = [S] + [ES] + [P]$. The concentrations of the different species can be normalized by the Michaelis constant $K_m = (k_{-1} + k_2)/k_1$ as $s = [S]/K_m$, $c = [ES]/K_m$, $e = [E]/K_m$ and $p = [P]/K_m$, and expressed as a function of the modified timescale $\theta = k_2 t$:

$$\left(1 - \frac{K_S}{K_m}\right)\frac{\mathrm{d}s}{\mathrm{d}\theta} = \frac{K_S}{K_m}c - es \tag{A2.2a}$$

$$\left(1 - \frac{K_S}{K_m}\right)\frac{\mathrm{d}c}{\mathrm{d}\theta} = es - c(1 + \Lambda) \tag{A2.2b}$$

$$\left(1 - \frac{K_S}{K_m}\right)\frac{\mathrm{d}e}{\mathrm{d}\theta} = c - es - \Lambda e \tag{A2.2c}$$

$$\frac{\mathrm{d}p}{\mathrm{d}\theta} = c \tag{A2.2d}$$

where $K_S = k_{-1}/k_1$ is the dissociation constant of the enzyme-substrate complex, and Λ is a normalization of λ:

$$\Lambda = \left(1 - \frac{K_S}{K_m}\right)\tau\lambda e_0 \tag{A2.3}$$

The system of ordinary differential equations comprising Eqs. A2.2a-A2.2d was numerically solved using the *ode15s* solver of Mathworks® MATLAB R2018a (Natick, MA, USA) for the sets of simulation parameters summarized in Tables A2.1, A2.3, and A2.4 (Section 2.8.5).

2.7.2. Irreversible enzyme inactivation - approximate analytical solution

The instantaneous concentration of total active enzyme (E_a) is given by the sum of active enzyme in its free (E) and bound state (ES). Since both forms are assumed to inactivate at the same rate, the evolution of E_a over time assumes the form of an exponential decay function:

$$E_a = E_0 e^{-\lambda t} \tag{A2.4}$$

For sufficiently high values of initial substrate concentration S_0, Pinto *et al.* (2015) [34] (see Introduction, Section II) showed that the reaction rate equation (Eq. A2.1d) can be rewritten as:

$$\frac{\mathrm{d}[P]}{\mathrm{d}t} = E_a \frac{S_0 - P}{S_0 - P + K_m} \tag{A2.5}$$

After replacing the time-dependent definition of E_a (Eq. A2.4), the analytical integration of A2.5 gives the evolution of $[P]$ as a function of the time t in the presence of enzyme inactivation:

$$\frac{[P]}{K_m} - \ln\left(\frac{S_0 - P}{S_0}\right) = \frac{1}{\tau\lambda}\left(1 - e^{-\lambda t}\right) \tag{A2.6}$$

with $\tau = K_m/(k_2 E_0)$. The limit of Eq. A2.6 for $t \to +\infty$ defines the final concentration of obtained product P_∞ as a function of S_0:

$$\frac{P_\infty}{K_m} - \ln\left(\frac{S_0 - P_\infty}{S_0}\right) = \frac{1}{\tau\lambda} \tag{A2.7}$$

The Lambert ω function is a built-in function in mathematical software that satisfies the transcendental equation $\omega(x)e^{\omega(x)} = x$. It is used to express Eqs. A2.6 and A2.7 as closed-form solutions, solving for $[P]$ (Eq. A2.8) and P_∞ (Eq. A2.9):

$$[P] = S_0 - K_m\omega\left[\frac{S_0}{K_m}\exp\left(\frac{S_0}{K_m} - \frac{1}{\tau\lambda}\left(1 - \exp\left(-\lambda t\right)\right)\right)\right] \tag{A2.8}$$

$$P_\infty = S_0 - K_m\omega\left[\frac{S_0}{K_m}\exp\left(\frac{S_0}{K_m} - \frac{1}{\tau\lambda}\right)\right] \tag{A2.9}$$

Eq. A2.9 was used to simulate P_∞/K_m vs S_0/K_m curves for the sets of simulation parameters summarized in Table A2.2 (Section 2.8.5).

2.7.3. Reaction rate analysis

Initial reaction rates (v_0) of progress curves for varying substrate concentrations were obtained by linear regression and analyzed to determine kinetic parameters K_m and $V = k_2 E_0$. Data were fitted to the MM equation (Eq. A2.10) by the non-linear least-squares method.

$$v_0 = \frac{V S_0}{S_0 + K_m} \tag{A2.10}$$

For procaspase-3 data, initial time intervals with approximate linear behavior were chosen for v_0 determination (Table A2.5, Section 2.8.5). For recombinant caspase-3 and α-thrombin data, initial velocity and instantaneous velocity values (v_0 and v_i, respectively) were determined. The selected time intervals used in each case are listed in Table A2.5 (Section 2.8.5). Instantaneous substrate concentrations S_i were determined for the median points of the intervals chosen for v_i determination. Non-linear least-squares fitting of v_i vs S_i was performed using the MM-like equation described by Pinto *et al.* (2015) [34]:

$$v_i = \frac{VS_i}{S_i + K_m} \tag{A2.11}$$

The fitted coefficients are presented with standard errors for 95% confidence level as estimated by the *fit* and *fitlm* functions of Mathworks® MATLAB 2018a (Natick, MA, USA).

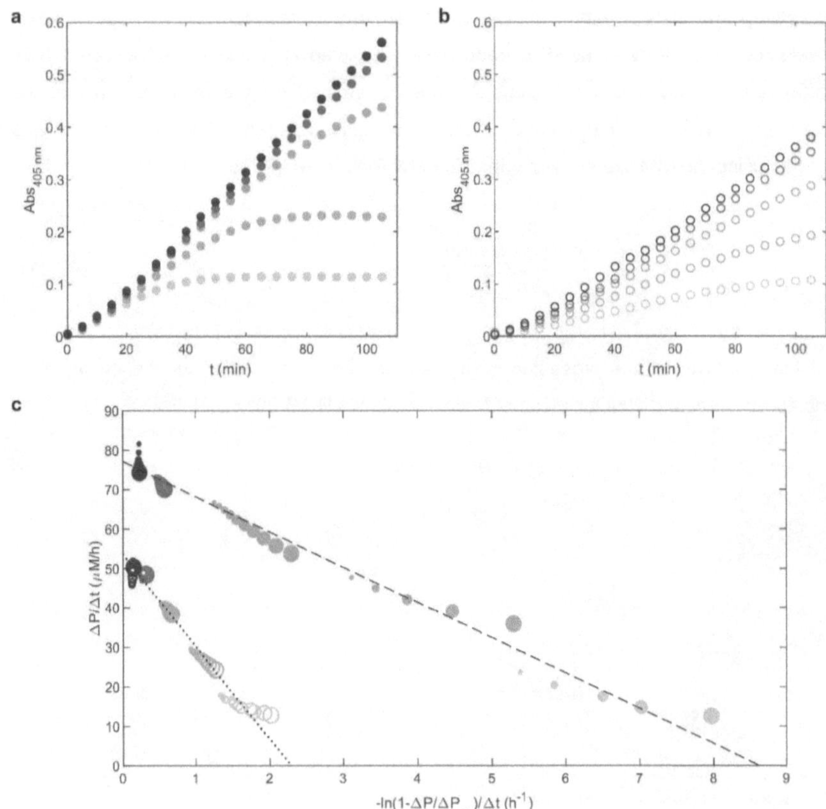

Figure A2.1. The presence of subnanomolar amounts (0.40 nM) of a synthetic variant of an anticoagulant produced by *D. andersoni* inhibits the catalysis of Tos-Gly-Pro-Arg-p-NA by 0.15 nM α-thrombin. (a and b) Absorbance (Abs_{405nm}) increase measured in the absence (a) and presence (b) of inhibitor for substrate concentrations of (from top to bottom) 400, 200, 100, 50 and 25 µM. (c) The same progress curves are represented according to the linearization method (LM) scale using $t_i = 30$ min (color-coded as in Figures A2.1a and A2.1b). Symbol size increases according to the time-course of the reaction. Lines: representation of the theoretical LM equation (Eq. 2.2) using $K_m^{app} = 8.94 \pm 0.17$ µM and $V^{app} = 77.1 \pm 0.5$ µM/h (dashed line, $R^2 = 0.9838$) and $K_m^{app} = 23.4 \pm 0.5$ µM and $V^{app} = 53.6 \pm 0.4$ µM/h (dotted line, $R^2 = 0.9730$).

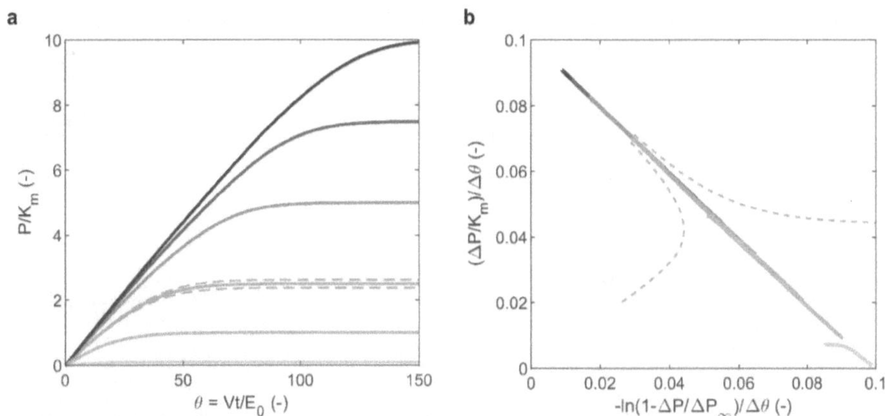

Figure A2.2. Small errors in reagent handling can be detected by the LM test. (a) Solid lines: theoretical progress curves calculated as described in Figure 2.2a for S_0/K_m values of (from top to bottom) 10, 7.5, 5, 2.5, 1, 0.1. Dashed lines: theoretical progress curves simulated for a random error of ±5% in the value of $S_0/K_m = 2.5$. (b) Lines: the same progress curves are represented in the LM scale using $t_i = 0$, and $P_\infty = S_0$. A clear deviation of the biased LM curves relatively to the overall trend can be identified since the beginning of the reaction.

2.8.5. Tables

Table A2.1. Model parameters used for simulation of reaction progress curves and LM curves presented in Figure 2.2.

$e_0 = E_0/K_m$ (-)	S_0/K_m (-)	K_s/K_m (-)	λ/k_2 (-)	$\tau\lambda = (\lambda/k_2)(1/e_0)$ (-)
	0.1			
	1			
	2.5			
0.1	5	0	0	0
	7.5			
	10			

Table A2.2. Model parameters used for simulation of P_∞/K_m vs S_0/K_m curves presented in Figure 2.6a.

$e_0 = E_0/K_m$ (-)	S_0/K_m (-)	K_s/K_m (-)	λ/k_2 (-)	$\tau\lambda = (\lambda/k_2)(1/e_0)$ (-)
			0	0
			10^{-6}	10^{-3}
			10^{-5}	0.01
			10^{-4}	0.1[a]
10^{-3}	[0,10]	1	1.5×10^{-4}	0.15
			2.5×10^{-4}	0.25
			5×10^{-4}	0.5
			1×10^{-3}	1
			5×10^{-3}	5

[a]Inactivation effects are poorly detected by the calibration-curve test for reference values of $\tau\lambda<0.1$.

Table A2.3. Model parameters used for simulation of reaction progress curves presented in Figure 2.6b.

$e_0 = E_0/K_m$ (-)	S_0/K_m (-)	K_s/K_m (-)	λ/k_2 (-)	$\tau\lambda = (\lambda/k_2)(1/e_0)$ (-)
10^{-3}				0.1[a]
1.3×10^{-3}				0.08
2×10^{-3}				0.05
4×10^{-3}	5	1	10^{-4}	0.03
0.01				0.01
0.1				10^{-3}

[a]Inactivation effects are poorly detected by the Selwyn test for reference values of $\tau\lambda<0.1$.

Table A2.4. Model parameters used for simulation of LM curves presented in Figure 2.6c.

$e_0 = E_0/K_m$ (-)	S_0/K_m (-)	K_s/K_m (-)	λ/k_2 (-)	$\tau\lambda = (\lambda/k_2)(1/e_0)$ (-)
	0.01			
	0.1			
	0.5			
	0.75			
10^{-3}	1	1	10^{-4}	0.1[a]
	1.5			
	2.5			
	5			
	10			

[a]Inactivation effects are still detected by the LM test for a reference value of $\tau\lambda=0.1$.

Table A2.5. Time intervals employed for determination of initial reaction velocity v_0 and instantaneous reaction velocity v_i by linear regression for procaspase-3, recombinant caspase-3 and α-thrombin enzymatic assays.

Procaspase-3	S_0 (µM)	3.125	6.25	12.5	25	50	100	150	200	300
	t_{int} (h) for v_0	[0,0.05]	[0,0.05]	[0,0.07]	[0,0.125]	[0,0.25]	[0,0.5]	[0,0.75]	[0,1]	[0,2]

Caspase-3	S_0 (µM)	3.125	6.25	12.5	25	50
	t_{int} (h) for v_0	[0,0.15]	[0,0.15]	[0,0.15]	[0,0.15]	[0,0.15]
	t_{int} (h) for v_i	[0.15,0.20]	[0.15,0.20]	[0.15,0.25]	[0.15,0.25]	[0.15,0.30]

α-thrombin $I = 0.40$ nM	S_0 (µM)	25	50	100	200	400
	t_{int} (min) for v_0	[0,30]	[0,30]	[0,30]	[0,30]	[0,30]
	t_{int} (min) for v_i	[30,60]	[30,60]	[30,60]	[30,60]	[30,60]

2.9. <u>Appendix B: Analysis of α-thrombin kinetics in the presence of two synthetic variants of an anticoagulant produced by *D. andersoni*</u>

The content of the present section shows an in-depth analysis of α-thrombin kinetics in the presence of two synthetic variants of an anticoagulant produced by <u>D. andersoni</u>, one of which was part of the work presented in the present chapter. This work was published in the following original research article:

*Watson, E.E., Ripoll-Rozada, J., Lee, A.C., Wu, M.C.L., Franck, C., Pasch T., Premdjee, B., Sayers, J., **Pinto, M.F.**, Martins, P.M., Jackson, S.P., Pereira, P.J.B., Payne, R.J. (2019), Rapid assembly and profiling of an anticoagulant sulfoprotein library. Proceedings of the National Academy of Sciences. 116(28):13873*

The synthetic inhibitor affecting the activity of α-thrombin and corresponding kinetic experiments were documented by Watson *et al.* (2019) [23]. The compound in question is synthetic andersonin 310 (And310), containing *O*-sulfotyrosine at positions 18 and 21 and N-methylated leucine and histidine at positions 41 and 44, respectively (And310 DS L41 H44); another variant of this anticoagulant was tested in this work possessing the same features except the absence of sulfatation (And310 Un L41 H44). Kinetic experiments were performed as described in Section 2.3.3, where modulator concentration ranges were of 0.1-3.2 nM and 31.25-1000 nM, for the DS and Un variants, correspondingly.

The sigmoid-shaped onsets observed for the obtained progress curves (Figures A2.3A and A2.3B) are admissible under conditions of high enzyme concentration that do not apply in the present context [34,36] (see Introduction, Section II, and Chapter 1, Section 1.4.2), or else for slow-onset inhibition due to enzyme isomerization [52, pp. 398-405]. Since increasing concentrations of inhibitor exacerbate the sigmoid shape, a mechanism in which the inhibitor binds to the productive, non-isomerized enzyme was proposed (Figures A2.4A and A2.4B) and numerically validated (Figures A2.3A, A2.3B and A2.4C). The sulfated homologue And310 DS L41 H44 was significantly more potent (K_i = 0.0738 ± 0.0136 nM) than And310 Un L41 H44 (K_i = 76.2 ± 32.2 nM), reflecting the importance of the tyrosine sulfate modifications for modulation of the thrombin inhibitory activity.

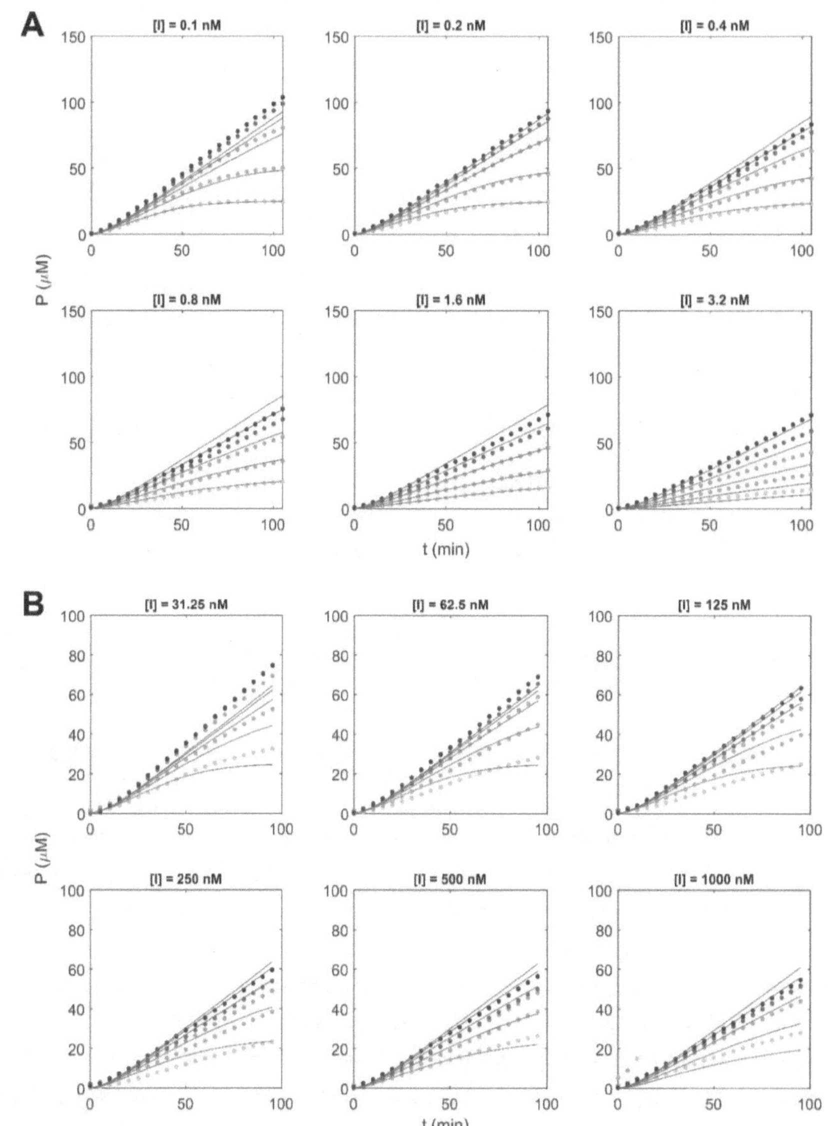

Figure A2.3. Kinetic analysis of the inhibitory activity of sulfated and unsulfated homologues of the doubly methylated variants of And310 against human α-thrombin for substrate concentrations of (symbols from top to bottom) 400, 200, 100, 50 and 25 µM. (A and B) Symbols: reaction product concentration measured over time for homologues (A) And310 DS L41 H44 and (B) And310 Un L41 H44; lines: numerical fit to the mechanism described in Figure A2.4.

A

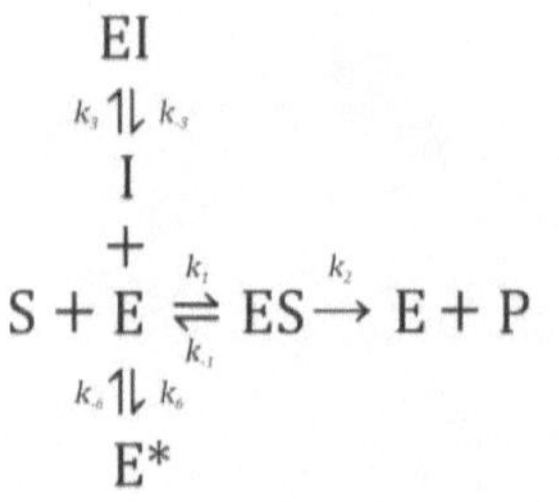

$$EI$$
$$k_3 \updownarrow k_{-3}$$
$$I$$
$$+$$
$$S + E \underset{k_{-1}}{\overset{k_1}{\rightleftharpoons}} ES \overset{k_2}{\rightarrow} E + P$$
$$k_{-6} \updownarrow k_6$$
$$E^*$$

B

$$\frac{d[S]}{dt} = k_{-1}[ES] - k_1[E][S]$$

$$\frac{d[P]}{dt} = k_2[ES]$$

$$\frac{d[E]}{dt} = (k_{-1} + k_2)[ES] - k_1[E][S] + k_{-6}[E^*]$$
$$- k_6[E] + k_{-3}[EI] - k_3[E][I]$$

$$\frac{d[ES]}{dt} = k_1[E][S] - (k_{-1} + k_2)[ES]$$

$$\frac{d[E^*]}{dt} = k_6[E] - k_{-6}[E^*]$$

$$\frac{d[EI]}{dt} = k_3[E][I] - k_{-3}[EI]$$

$$\frac{d[I]}{dt} = -k_3[E][I] + k_{-3}[EI]$$

C

	And310 DS L41 H44	And310 Un L41 H44
k_1 ($\mu M^{-1} min^{-1}$)	$4.55 \times 10^3 \pm 2.02 \times 10^4$	$9.30 \times 10^3 \pm 1.04 \times 10^6$
k_{-1} (min^{-1})	$8.66 \times 10^3 \pm 6.93 \times 10^4$	$1.00 \times 10^4 \pm 1.70 \times 10^6$
k_2 (min^{-1})	$6.75 \times 10^3 \pm 89.7$	$5.17 \times 10^3 \pm 95.0$
k_6 (min^{-1})	0.219 ± 0.080	0.546 ± 0.195
k_{-6} (min^{-1})	0.0805 ± 0.0096	0.0892 ± 0.098
k_3 ($\mu M^{-1} min^{-1}$)	$2.11 \times 10^5 \pm 1.08 \times 10^4$	863 ± 308
k_{-3} (min^{-1})	15.6 ± 2.75	65.7 ± 15.0
K_I (nM)	0.0738 ± 0.0136	76.2 ± 32.2

Figure A2.4. Slow-onset inhibition due to enzyme isomerization. (A) Schematic mechanism, where E and E^* represent the non-isomerized and isomerized enzyme; S, I and P represent substrate, inhibitor and product; ES and EI represent the enzyme-substrate and enzyme-inhibitor complexes. (B) System of ordinary differential equations describing the proposed mechanism; square brackets represent molar concentrations. The solver *ode15s* of Mathworks® MATLAB R2018a (Natick, MA, USA) was used to numerically solve this system in two steps: (1) the equilibrium concentrations of E and E^* are obtained using as total $[E] = 0.30$ nM and $[S] = [I] = 0$ (pre-mixing conditions); (2) the system is solved using as initial conditions the concentrations of S and I corresponding to each assay and the halved equilibrium concentrations of E and E^* (reaction start conditions). (C) Fitting results obtained using function *lsqcurvefit* of Mathworks® MATLAB R2018a (Natick, MA, USA), where results are presented as fitted parameters ± standard errors for 95% confidence level.

References

1. Holdgate, G.A., Meek, T.D., and Grimley, R.L. (2017), *Mechanistic enzymology in drug discovery: a fresh perspective.* Nature Reviews: Drug Discovery. 17:115.

2. Rask-Andersen, M., Masuram, S., and Schiöth, H.B. (2014), *The Druggable Genome: Evaluation of Drug Targets in Clinical Trials Suggests Major Shifts in Molecular Class and Indication.* Annual Review of Pharmacology and Toxicology. 54(1):9-26.

3. Thorne, N., Auld, D.S., and Inglese, J. (2010), *Apparent activity in high-throughput screening: origins of compound-dependent assay interference.* Current Opinion in Chemical Biology. 14(3):315-324.

4. Aldrich, C., *et al.* (2017), *The Ecstasy and Agony of Assay Interference Compounds.* ACS Central Science. 3(3):143-147.

5. Selwyn, M.J. (1965), *A simple test for inactivation of an enzyme during assay.* Biochimica et Biophysica Acta (BBA) - Enzymology and Biological Oxidation. 105(1):193-195.

6. Schnell, S. and Hanson, S.M. (2007), *A test for measuring the effects of enzyme inactivation.* Biophys Chem. 125(2-3):269-74.

7. Feng, B.Y. and Shoichet, B.K. (2006), *A detergent-based assay for the detection of promiscuous inhibitors.* Nat Protoc. 1(2):550-3.

8. Rexer, T.F.T., *et al.* (2018), *One pot synthesis of GDP-mannose by a multi-enzyme cascade for enzymatic assembly of lipid-linked oligosaccharides.* Biotechnology and Bioengineering. 115(1):192-205.

9. Yu, K., *et al.* (2011), *A high-throughput colorimetric assay to measure the activity of glutamate decarboxylase.* Enzyme and Microbial Technology. 49(3):272-276.

10. Cao, W. and De La Cruz, E.M. (2013), *Quantitative full time course analysis of nonlinear enzyme cycling kinetics.* Scientific Reports. 3:2658.

11. Shoichet, B.K. (2006), *Screening in a spirit haunted world.* Drug Discovery Today. 11(13):607-615.

12. Grosch, J.-H., *et al.* (2017), *Influence of the experimental setup on the determination of enzyme kinetic parameters.* Biotechnology Progress. 33(1):87-95.

13. Gielen, F., *et al.* (2013), *A Fully Unsupervised Compartment-on-Demand Platform for Precise Nanoliter Assays of Time-Dependent Steady-State Enzyme Kinetics and Inhibition.* Analytical Chemistry. 85(9):4761-4769.

14. Cornish-Bowden, A. (2015), *One hundred years of Michaelis–Menten kinetics.* Perspectives in Science. 4:3-9.

15. Rawlings, N.D., *et al.* (2018), *The MEROPS database of proteolytic enzymes, their substrates and inhibitors in 2017 and a comparison with peptidases in the PANTHER database.* Nucleic Acids Res. 46(D1):D624-d632.

16.   Boatright, K.M. and Salvesen, G.S. (2003), *Mechanisms of caspase activation*. Current Opinion in Cell Biology. 15(6):725-731.

17.   Crawley, J.T.B., *et al.* (2007), *The central role of thrombin in hemostasis*. Journal of Thrombosis and Haemostasis. 5(s1):95-101.

18.   Corral-Rodríguez, M.Á., *et al.* (2009), *Tick-derived Kunitz-type inhibitors as antihemostatic factors*. Insect Biochemistry and Molecular Biology. 39(9):579-595.

19.   Corral-Rodríguez, M.Á., *et al.* (2010), *Leech-Derived Thrombin Inhibitors: From Structures to Mechanisms to Clinical Applications*. Journal of Medicinal Chemistry. 53(10):3847-3861.

20.   Ware, F.L. and Luck, M.R. (2017), *Evolution of salivary secretions in haematophagous animals*. Bioscience Horizons. 10:hzw015.

21.   Parizi, L.F., *et al.* (2018), *Peptidase inhibitors in tick physiology*. Med Vet Entomol. 32(2):129-144.

22.   Francischetti, I.M., *et al.* (2009), *The role of saliva in tick feeding*. Front Biosci (Landmark Ed). 14:2051-88.

23.   Watson, E.E., *et al.* (2019), *Rapid assembly and profiling of an anticoagulant sulfoprotein library*. Proceedings of the National Academy of Sciences. 116(28):13873.

24.   Pereira, C., *et al.* (2014), *Potential small-molecule activators of caspase-7 identified using yeast-based caspase-3 and -7 screening assays*. Eur J Pharm Sci. 54:8-16.

25.   Gloria, P.M.C., *et al.* (2011), *Aspartic vinyl sulfones: inhibitors of a caspase-3-dependent pathway*. Eur J Med Chem. 46(6):2141-6.

26.   Boucher, D., Duclos, C., and Denault, J.-B., (2014), Caspases, Paracaspases and Metacaspases: Methods and Protocols, ed. P.V. Bozhkov and G. Salvesen. Humana Press, Chp.

27.   Thompson, R.E., *et al.* (2017), *Tyrosine sulfation modulates activity of tick-derived thrombin inhibitors*. Nature Chemistry. 9:909.

28.   Schnell, S. and Mendoza, C. (1997), *Closed Form Solution for Time-dependent Enzyme Kinetics*. Journal of Theoretical Biology. 187(2):207-212.

29.   Duggleby, R.G. (2001), *Quantitative analysis of the time courses of enzyme-catalyzed reactions*. Methods. 24(2):168-174.

30.   Orsi, B.A. and Tipton, K.F. (1979), *Kinetic analysis of progress curves*. Methods in Enzymology. 63:159-83.

31.   Eicher, J., Snoep, J., and Rohwer, J. (2012), *Determining Enzyme Kinetics for Systems Biology with Nuclear Magnetic Resonance Spectroscopy*. Metabolites. 2(4):818.

32.   Stroberg, W. and Schnell, S. (2016), *On the estimation errors of K_M and V from time-course experiments using the Michaelis–Menten equation*. Biophysical Chemistry. 219:17-27.

33. Hanson, S.M. and Schnell, S. (2008), *Reactant Stationary Approximation in Enzyme Kinetics*. The Journal of Physical Chemistry A. 112(37):8654-8658.

34. Pinto, M.F., *et al.* (2015), *Enzyme kinetics: the whole picture reveals hidden meanings*. FEBS Journal. 282(12):2309-2316.

35. Walker, A.C. and Schmidt, C.L.A. (1944), *Studies on histidase*. Archives of Biochemistry and Biophysics. 5:445-467.

36. Pinto, M.F. and Martins, P.M. (2016), *In search of lost time constants and of non-Michaelis-Menten parameters*. Perspectives in Science. 9:8-16.

37. Fleisher, G.A. (1953), *Curve Fitting of Enzymatic Reactions Based on the Michaelis—Menten Equation[1]*. Journal of the American Chemical Society. 75(18):4487-4490.

38. Briggs, G.E. and Haldane, J.B.S. (1925), *A note on the kinetics of enzyme action*. Biochemical Journal. 19:338-339.

39. Bisswanger, H. (2014), *Enzyme assays*. Perspectives in Science. 1(1–6):41-55.

40. Tang, Q. and Leyh, T.S. (2010), *Precise, Facile Initial Rate Measurements*. The Journal of Physical Chemistry B. 114(49):16131-16136.

41. Bose, K., *et al.* (2003), *An Uncleavable Procaspase-3 Mutant Has a Lower Catalytic Efficiency but an Active Site Similar to That of Mature Caspase-3*. Biochemistry. 42(42):12298-12310.

42. Cornish-Bowden, A. (1975), *The use of the direct linear plot for determining initial velocities*. Biochemical Journal. 149(2):305-312.

43. Cornish-Bowden, A., (2012), *Fundamentals of Enzyme Kinetics*, 4th ed. Wiley-Blackwell (Weinheim, Germany), pp.44.

44. Matosevic, S., Szita, N., and Baganz, F. (2011), *Fundamentals and applications of immobilized microfluidic enzymatic reactors*. Journal of Chemical Technology & Biotechnology. 86(3):325-334.

45. Sørensen, T.H., *et al.* (2015), *Temperature effects on kinetic parameters and substrate affinity of Cel7A cellobiohydrolases*. Journal of Biological Chemistry.

46. Schnell, S. (2014), *Validity of the Michaelis-Menten equation – steady-state or reactant stationary assumption: that is the question*. The FEBS Journal. 281(2):464-472.

47. Botts, J. and Morales, M. (1953), *Analytical description of the effects of modifiers and of enzyme multivalency upon the steady state catalyzed reaction rate*. Transactions of the Faraday Society. 49(0):696-707.

48. Henderson, P.J.F. (1973), *Steady-state enzyme kinetics with high-affinity substrates or inhibitors. A statistical treatment of dose–response curves*. Biochemical Journal. 135(1):101.

49. Morrison, J.F. (1969), *Kinetics of the reversible inhibition of enzyme-catalysed reactions by tight-binding inhibitors.* Biochimica et Biophysica Acta (BBA) - Enzymology. 185(2):269-286.

50. Szedlacsek, S.E. and Duggleby, R.G., (1995), [6] Kinetics of slow and tight-binding inhibitors. Vol. 249. Academic Press, Chp.

51. Ben Halima, S., *et al.* (2016), *Specific Inhibition of β-Secretase Processing of the Alzheimer Disease Amyloid Precursor Protein.* Cell Reports. 14(9):2127-2141.

52. Baici, A., (2015), *Kinetics of Enzyme-Modifier Interactions*, 1st ed. Springer.

53. Swainston, N., *et al.* (2018), *STRENDA DB: enabling the validation and sharing of enzyme kinetics data.* FEBS Journal. 285(12):2193-2204.

54. Buchholz, P.C.F., *et al.* (2016), *BioCatNet: A Database System for the Integration of Enzyme Sequences and Biocatalytic Experiments.* ChemBioChem. 17(21):2093-2098.

Chapter 3.

interferENZY: a web-based tool for enzymatic assay validation and standardized kinetic analysis

The content of the present chapter constitutes an original research article in preparation (Article III):

Pinto, M.F., Baici, A., Pereira, P.J.B., Macedo-Ribeiro, S., Pastore, A., Rocha, F., Martins, P.M., interferENZY: a web-based tool for enzymatic assay validation and standardized kinetic analysis.

3.2. <u>Introduction</u>

Variations in enzyme activity induced by structural changes or by the presence of small-molecule ligands are quantitatively characterized through enzyme kinetics analysis. Adequate kinetic data interpretation is therefore essential to ensure the progress of fundamental and applied research in disciplines such as enzyme engineering, systems biology and drug discovery [2]. An effort toward the standardization of data reporting procedures has been undertaken over the last few years, e.g., with the creation of the Standards for Reporting Enzyme Data (STRENDA) Consortium [3] and of the database STRENDA DB [2]. In addition, there is a strong urge for systematized methods of enzymatic assay validation and kinetic parameter estimation as illustrated by the availability of software and web-based tools for determining kinetic parameters by initial velocity and/or full progress curve analysis, including computer programs *FITSIM/KINSIM* [4,5], *DYNAFIT* [6] and *KinTek Global Kinetic Explorer* [7], computational packages for programming platforms R [8] and MATLAB (e.g., *PCAT* [9]), and web-based tools including *ENZO* [10] and the *Continuous Enzyme Kinetics Analysis Tool* [11]. While these programs constitute different alternatives to numerically fit known reaction mechanisms to selected kinetic data, limited insight is provided on the reliability of the experimental bioassays used for data collection. Unsuspected interferences such as those caused by enzyme inactivation and inhibition are known to have major effects on the reproducibility of kinetic results [1,12] (see Chapter 2, Sections 2.5.1 and 2.5.4). These effects can substantially accumulate over time, thereby affecting the determination of kinetic parameters by full time-course analysis [13,14]. On the other hand, classical initial velocity analysis can heavily depend on subjective selection criteria, especially under conditions of low substrate concentration for which the "initial linear phase" is not easily observable [15]. Fast enzyme kinetics are particularly troublesome, since the important moments of the reaction take place during the lag period preceding effective reaction monitoring [1,16] (see Chapter 2, Section 2.5.3).

With the goal of detecting hidden disturbances affecting enzymatic assays, a simple linearization method (LM) was recently proposed based on the representation of reaction progress curves in terms of the following X and $\bar{Y}$ coordinates [1] (see Chapter 2, Section 2.4):

$$X = -\frac{\ln(1 - \Delta P/\Delta P_\infty)}{\Delta t} \tag{3.1a}$$

$$Y = \frac{\Delta P}{\Delta t} \tag{3.1b}$$

where ΔP is the product concentration increase measured over a period of time Δt, and ΔP_∞ refers to the product concentration increase observed at the end of the time-course. The LM is highly sensitive to assay interferences even when their presence is not suspected *a priori*. Its application does not require carrying out additional experiments and provides unbiased estimations of the apparent kinetic parameters K_m^{app} and V^{app}; in the absence of interferences, these correspond to the Michaelis constant (K_m) and to the limiting rate (V) [13,14,17], respectively [1].

The robustness and general interest of the LM motivated its implementation as a free web-tool available for automatic quality control of enzymatic assays and for parameter estimation purposes. Here we present the webserver interferENZY (*INTERFERences + ENZYmology*) to (i) automatically process user-provided experimental data, (ii) evaluate the quality of the tested enzymatic assay and, for successfully validated cases, (iii) provide impartial estimates of K_m^{app} and V^{app} parameters. Each of these features is illustrated using practical examples corresponding to the study cases documented in Chapter 2, and a case obtained for the model enzymatic system evaluated in Chapter 4, that, taken together, serve to demonstrate interferENZY's usefulness in improving the quality of enzymology data analysis.

3.3. <u>Methods</u>

3.3.1. Webserver infrastructure

The interferENZY webserver was built on a Linux-based machine (CentOS 7.6) running GNU Octave CLI 5.1.0 (free software under the GNU General Public License) with the function packages *io*, *statistics*, and *image*, and using the *gnuplot* toolkit for graphical outputs. Data analysis by interferENZY is performed dynamically by a custom script written according to the webserver pipeline detailed in Figure 3.1.

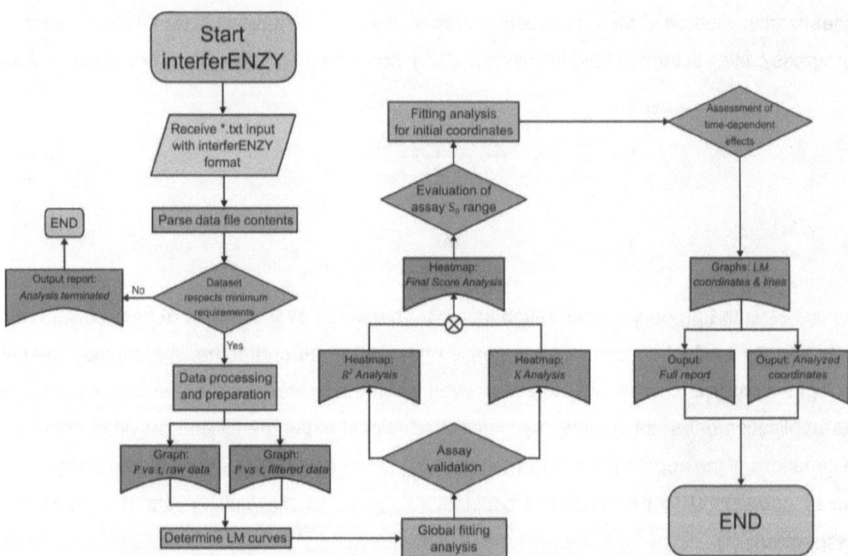

Figure 3.1. Summarized version of the interferENZY webserver pipeline (full flowchart in Figure A3.1, Appendix).

3.4. <u>Data input and parsing</u>

Data are input in the interferENZY webpage interface (Figure 3.2A) by uploading an individual tab-delimited file (*.txt) containing a kinetic dataset (mandatory format described in Figure A3.2, Appendix). For convenience, a Microsoft Excel spreadsheet template is provided, which can optionally be filled and saved as tab-delimited text (*.txt format). A pre-filled example containing a set of numerically simulated progress curves is also provided for reference. The interferENZY webserver can analyze reaction time-courses of (i) product production or substrate decay (the latter are converted to product production curves); (ii) continuous or stopped assay measurements; (iii) variable measurement frequencies; (iv) variable assay duration. Datasets analyzed by interferENZY must comprise a minimum of 5 time-courses obtained at 5 different concentrations of substrate (S_0), with fixed concentration of enzyme (E_0), in the same concentration units as S_0; each time-course must contain a minimum of two reaction points. If any of these requirements is not met, dataset analysis is interrupted and the reason for interruption is identified in the output report.

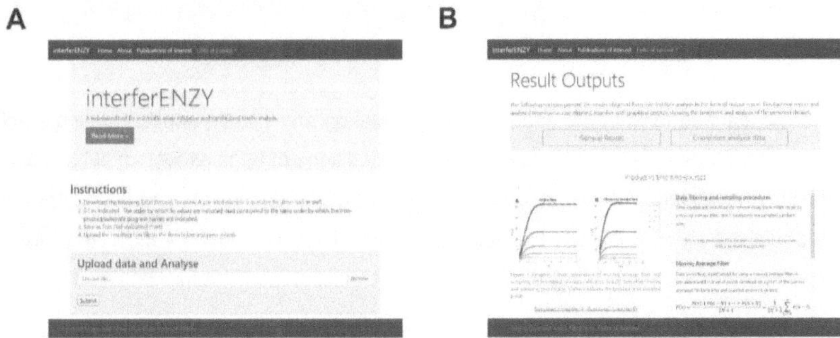

Figure 3.2. interferENZY user interface. (A) Main webpage and interface for dataset submission. (B) Result output page. Download links are provided for the output reports and high-resolution versions of the graph previews. A short result interpretation guide is provided next to each graphical output.

3.4.1. Linearization Method (LM) processing

Following data input and parsing, the number of time points in each curve is quantified. Only datasets with progress curves with at least 5 time points are subjected to curve filtering and point sampling (Figure A3.1, Appendix).

Data smoothing is performed using a moving average filter, followed by the determination of 7 evenly spaced concentrations determined based on each curve's maximum and minimum concentrations (product-wise selection). The points closest to each estimated product concentration are then sampled with the corresponding time instants (point sampling). If all reactions are assumed to reach completion (i.e., P_∞ values for all time-courses correspond to ±10% of the respective substrate concentrations), Eqs. 3.1a and 3.1b are used to compute the LM coordinates. If this internal criterion is not met, the values of X are computed using a modified version of Eq. 3.1a in which ΔP_∞ is replaced by the corresponding predicted value $\Delta S_0 = S_0 - P_i$. Whenever allowed by the total number of sampled points, a maximum of 3 different scenarios are considered assuming 3 different initial conditions. Postponed values of t_i are used in order to circumvent any altered behavior in the initial stages of the reaction arising from the stabilization of reaction conditions or from complex enzymatic mechanisms [1] (see Chapter 2, Sections 2.5.3 and 2.5.4).

Parameters K_m^{app} and V^{app} are determined for each t_i scenario using the LM coordinates obtained for all substrate concentrations. The set of (X,Y) coordinates are numerically fitted by the LM equation,

$$Y = V^{app} - K_m^{app}X \tag{3.2}$$

which is a linearized form of the integrated MM equation and can be applied to single active site, single substrate catalytic mechanisms under conditions of large excess of substrate over enzyme. In the absence of assay interferences, progress curves obtained for different S_0 values and fixed E_0 produce superimposable straight lines with slope $-K_m^{app}$ and intercept V^{app}.

3.4.2. Result interface & Output scores

Results from interferENZY analysis (displayed in a new page; Figure 3.2B) contain progress curve graphs (raw data and filtered/sampled data), LM coordinates graphs for each tested t_i scenario, and heatmaps for assay score evaluation. Short guides for result interpretation are offered next to each set of graphical results. Download links are provided for high-quality versions of the graphs. An output report (text format) is generated containing the data treatment steps, output parameters and statistical analysis, and comments/warnings on enzymatic assay quality. A second text file containing the filtered and sampled progress curves, as well as modified reaction coordinates resulting from interferENZY analysis is also generated. Both files are available for download.

Results from statistical analysis of the linear regressions for each studied scenario include the standard errors for each kinetic parameter (95% confidence), the coefficient of determination (R^2), F and p values, and the estimated error variance. Individual curve trends (for each assayed S_0) are evaluated by the Kendall (K) correlation coefficient for datasets containing progress curves with at least 5 time points.

The quality of the enzymatic assay is assessed by the values of R^2 and K. A score of 0 (lowest) to 20 (highest) is attributed to each of these parameters, considering their possible range (0 to 1 for R^2, and 1 to −1 for K). The overall quality of the enzymatic assay is characterized by a final score obtained by the weighted average of R^2 scores and the average of K scores for all the individual curves (75% and 25% weight, respectively). The distributions of R^2 values, K values, and the final scores obtained are represented as heatmaps for each t_i scenario and, in the case of coefficient K, for each S_0 value. Each score is converted to a color in a red-to-white-to-blue gradient, where red, white and blue correspond to score values of 0, 10 and 20, respectively.

3.5. <u>Results and Discussion</u>

3.5.1. Data filtering and sampling

The application of moving average filters to the input raw data decreases signal noise and facilitates the LM analysis. On the other hand, excessive data usage is avoided by sampling the catalytic reaction only at representative moments of the time-course. These procedures are illustrated for a dataset obtained by measuring the activity of hen egg-white lysozyme on the fluorogenic substrate MUF-triNAG (Figure 3.3).

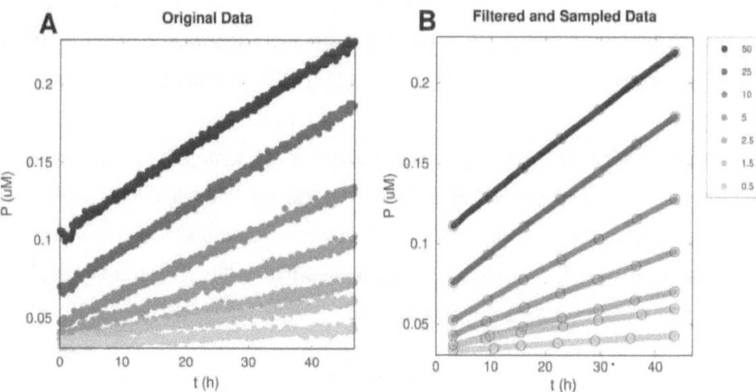

Figure 3.3. Data filtering and sampling by interferENZY. Reaction progress curves of MUF-triNAG ($S_0 = [0.5, 50]$ µM) catalysis by lysozyme ($E_0 = 0.25$ µM) [18], (A) before and (B) after data processing. Red markers indicate product-wise sampled points.

Lysozyme is reputedly a slow enzyme, requiring long periods of time to produce assay readouts substantially above the baseline [18]. Contrary to initial rate analysis, the application of the LM is not limited to a specific reaction phase, thereby allowing signal noise reduction using numerical filters (Figure 3.3A and Figure 3.3B).

Useful for analyzing progress curves obtained in separate experiments, the dynamic parsing algorithm allows different measuring frequencies and/or assay durations. In the lysozyme example (Figure 3.3), reaction completion would require unrealistically long measuring times, so the values of P_∞ are estimated assuming exact correspondence with S_0. It is therefore important that the calibrated readouts are provided in the same concentration units as the initial substrate concentrations.

interferENZY analysis is compatible with the usual progress curves measured by quantifying the concentration of product over time, but it can also be used for analysis of substrate decay curves, for which the approximation $P \approx S_0 - S$ is employed. This is applicable in conditions of large excess of substrate over enzyme ($E_0 \ll S_0 + K_m$) that are also required for the validity of the steady-state approximation (SSA) [1,19] (see Introduction, Section I.I, and Chapter 1, Section 1.5), and of the LM analysis [1] (see Chapter 2, Section 2.4). The oxidation of NADH to NAD$^+$ by enzymes such as NADH dehydrogenases [20] or lactate dehydrogenase [21] is a typical example of interest where it is the substrate (NADH) and not the product (NAD$^+$) that is quantified over time [22].

The interferENZY webserver requires a minimum of 5 assayed values of S_0, ideally covering a range between $\sim 0.1 K_m^{app}$ and $\sim 10 K_m^{app}$ that includes all regions of the v_0 vs S_0 curve [23]. Covering this range is not always possible due to either low substrate solubility or optical assay limitations (e.g., spectrophotometry detection ranges) [23]. In addition, more than 20% of all known enzymes are inhibited by their substrate(s) [24], which can produce deviations from the expected kinetic behavior [25]. Each progress curve should comprise at least two time points, which are used (along with the values of S_0 or P_∞) to compute the linearized coordinates (Eqs. 3.1a,3.1b). Stopped (or discontinuous) assays, in which discrete concentration measurements are generally preceded by reaction quenching [23,26], should therefore include a minimum of two time-separated quantifications.

3.5.2. Automatic LM application

The diagnostics of assay interferences is exemplified for the catalysis of the fluorogenic substrate Ac-DEVD-AMC by procaspase-3, obtained from yeast cell extracts (Figure 3.4) [1] (see Chapter 2, Sections 2.5.1 and 2.5.2).

Even though the observed progress curves do not show any evident deviation from the expected exponential trend (Figure 3.4A, 3.4B), the pattern defined by the corresponding linearized coordinates is clearly anomalous: non-superimposing curves with positive slopes (rather than negatively-sloped, superimposing straight lines) suggest enzyme inactivation (Figure 3.4C-3.4E). In fact, the rapid loss of procaspase-3 activity has been confirmed by other indirect methods and by direct measurements of its activity over time [1] (see Chapter 2,

Section 2.5.2). Exceedingly prolonged incubation times (Figure 3.4A, 3.4B) can lead to kinetic-altering events such as enzyme inactivation, substrate depletion, substrate inhibition, and product inhibition [27, pp.45-59]. The three scenarios analyzed for procaspase-3 (Figure 3.4C-3.4E) were generated assuming three different instants – t_1, t_2 and t_3 – as initial condition of the LM.

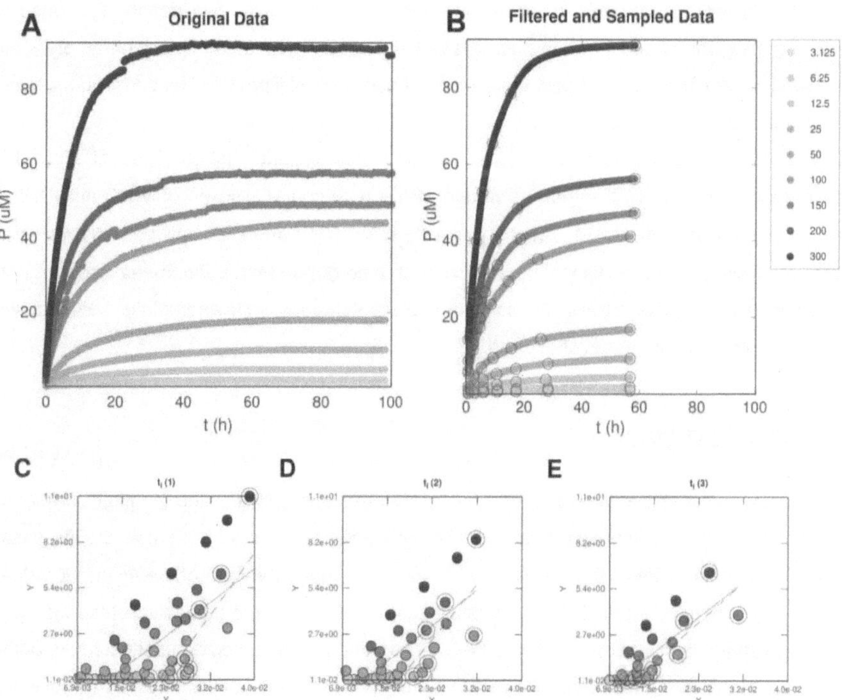

Figure 3.4. Example of automatic application of the LM to published datasets. (A,B) Progress curves of Ac-DEVD-AMC ($S_0 = [3.125, 300]$ µM) catalysis by procaspase-3 ($E_0 = 0.123$ mg/mL [1], Chapter 2, Sections 2.5.1 and 2.5.2; (A) before and (B) after data processing (red markers: product-wise sampled points). (C-E) Symbols: linearized coordinates obtained using (C) $i = 1$, (D) $i = 2$, and (E) $i = 3$ as the LM first instant (t_i); open symbols refer to initial coordinates of each reaction. Lines: linear regressions obtained using (red) all or (blue) initial-only LM coordinates.

As shown in the following examples, this practice can improve the quality of kinetic results by minimizing the impact of random experimental errors occurring on the reaction onset. In the case of procaspase-3, none of the three scenarios was appropriate for kinetic parameter estimation because interferences caused by enzyme inactivation increase with time becoming more pronounced at later reaction stages. The fitted kinetic parameters are initially evaluated in terms of their physical meaning, with negative K_m^{app} or V^{app} values, or V^{app} values lower than the lowest Y coordinate being sufficient reasons for assay invalidation. The ensuing warning message is added to the output report wherein only validated kinetic parameters are included. When failing the LM test, auxiliary diagnostics are still provided for subsequent assay troubleshooting and improvement.

The graphical representations of (X, Y) coordinates include straight lines fitted to LM coordinates covering (i) the entire duration of the time-course and (ii) only the initial (X,Y) coordinate (Figure 3.4C-3.4E). Variations $> 10\%$ in R^2, K_m^{app} and V^{app} between full and initial-only analysis are an additional indication that time-dependent phenomena are possibly affecting the progress curves. Portions of the time-courses with anomalous behavior will markedly deviate from the overall trends.

3.6. Quality control

Two statistical coefficients, the coefficient of determination R^2 and the Kendall correlation coefficient K, and a composed "final score" are provided to summarily characterize the assay quality. The interferENZY final score is a 75%/25% weighted average of the R^2 and K values previously converted into a 0-20 quantitative scale. One low score dataset, corresponding to the procaspase-3 progress curves (Figure 3.4), and one well-behaved dataset obtained for the catalysis of the same substrate (Ac-DEVD-AMC) by caspase-3, are exemplified (Figure 3.5, and Figure A3.3 in Appendix). The resulting statistical indicators are represented in a red-to-white-to-blue gradient heatmap ranging from bad (red) to average (white) to good (blue) scores for each considered scenario. The analysis of different initial instants t_i improved the R^2 evaluation (Figure 3.5D) and the final score (Figure 3.5F) of the caspase-3 assay, which is known to be affected by an initial period of ~10 min of temperature stabilization [1] (see Chapter 2, Section 2.5.3).

Illustrating the advantage of the combined analysis of different correlation estimators, the quality of the procaspase-3 assay would not be fully disclosed if only coefficients of determination were evaluated (Figure 3.5A-3.5C). Unlike the R^2 coefficient, which accounts for all time points of all progress curves, the K coefficient is determined for each time-course and is used to express monotonicity rather than linearity of rank correlations [28,29]. For successfully validated assays such as that of caspase-3 (Figure 3.5D-3.5E), the top-scored t_i scenario is selected to provide the best estimate of kinetic parameters given in the interferENZY output report. Warning messages are produced if the value of K_m^{app} falls outside the adopted S_0 range, if the highest S_0 concentration is $< 10K_m^{app}$ or if the lowest S_0 concentration is $> 0.1K_m^{app}$ [23].

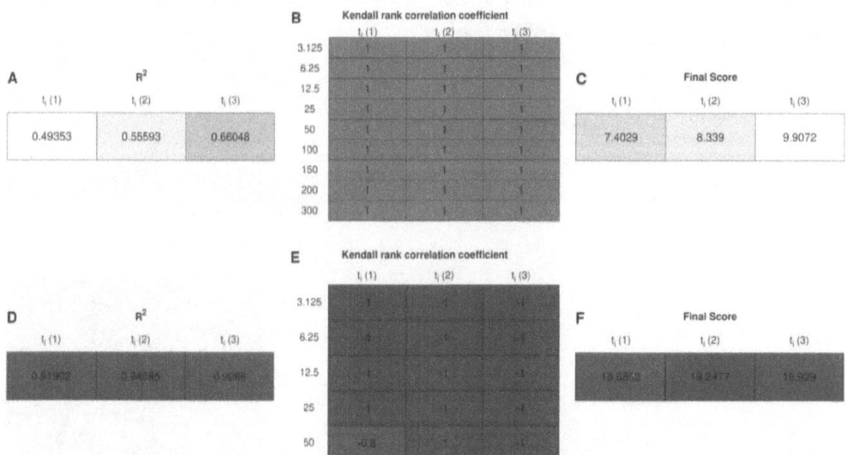

Figure 3.5. Quality control analyses using two published datasets. Statistical outputs of interferENZY for (A-C) the procaspase-3 dataset (Figure 3.4) and for (D-E) Ac-DEVD-AMC ($S_0 = [3.125, 50]$ µM) catalysis by caspase-3 ($E_0 = 1$ U [1]), Chapter 2, Sections 2.5.1, 2.5.2 and 2.5.3. Heatmap representations of (A,D) the coefficient of determination R^2, (B,E) the Kendall correlation coefficient, and (C,F) the interferENZY final score for the different t_i scenarios.

Underperforming assays with lower than expected interferENZY scores can be explained by factors not directly linked with the occurrence of extraneous interferences [1]. The catalysis of the chromogenic substrate Tos-Gly-Pro-Arg-p-nitroanilide by α-thrombin in the presence of an anticoagulant produced by *Dermacentor andersoni* is, in this respect, an interesting case, as it involves enzyme isomerization steps that are not present in the Briggs-Haldane mechanism upon which the LM is based [1,30] (see Chapter 2, Sections 2.5.4 and 2.9). The sigmoid-shaped onset of these progress curves (Figure 3.6A) is exacerbated for higher anticoagulant concentrations, which configures a possible mechanism of inhibitor binding to productive, non-isomerized enzyme [30]. Despite the nearly invariant R^2 values (Figure 3.6B), the K-based diagnostics and (consequently) the interferENZY final score markedly improve as the initial instant t_i is successively postponed (Figure 3.6C, 3.6D). The expected linear trend of LM-transformed time-courses is, however, not fully recovered even when the initial instant is set to >30 min after reaction monitoring is started (Figure A3.4, Appendix). Hence, the changes in interferENZY scores are more likely explained by time-dependent modulation effects rather than unrealistically long periods of experimental conditions stabilization. Those effects, which in the current didactic example are deliberately provoked by the addition of α-thrombin inhibitors, can in practice occur due to the presence of vestigial amounts of high-affinity modifiers [1] (see Chapter 2, Section 2.5.4).

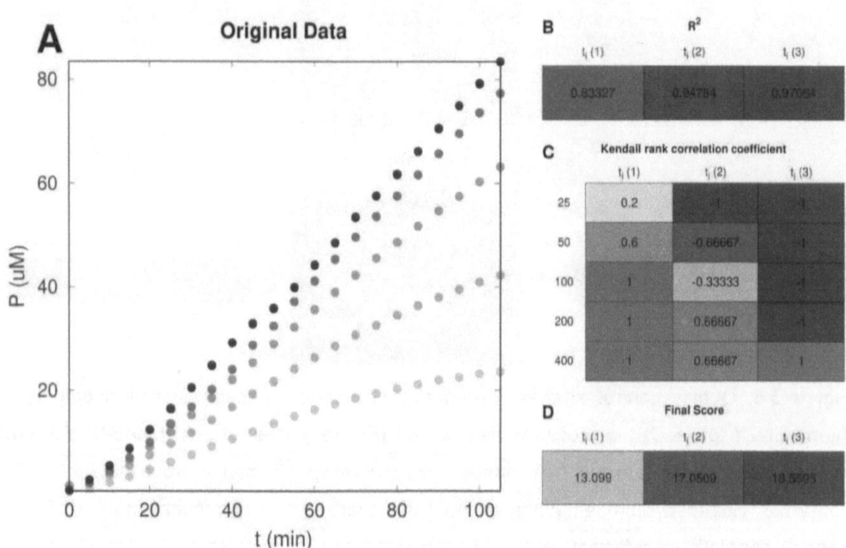

Figure 3.6. Time-dependent enzyme modulation mechanisms can be detected by interferENZY. (A) Progress curves of Tos-Gly-Pro-Arg-p-nitroanilide catalysis ($S_0 = 400, 200, 100, 50$ and 25 µM from top to bottom) by α-thrombin ($E_0 = 0.15$ nM) in the presence of 0.40 nM inhibitor [1,30] (Chapter 2, Sections 2.5.4 and 2.9). (B-D) Statistical outputs of interferENZY represented as heatmaps of (B) the coefficient of determination R^2, (C) the Kendall correlation coefficient, and (D) the interferENZY final score for the different t_i scenarios.

When less than 5 time points are available per progress curve, individual rank correlation coefficients are not determined ($K = 0$), so the maximum possible final score for interferENZY analysis becomes 15. This automatic penalty is implemented because ill-defined individual trends are not suitable for complete LM analysis. This is the case of stopped assays, where few time points per time-course are generally collected. Although kinetic parameter estimation is still possible for these assays, the breadth of the quality-control reports is evidently limited by the amount of data that is provided. In the application example of L-cysteine catalysis by the desulfurase IscS in the presence of the scaffold protein IscU (see Chapter 4, Sections 4.2.1 and 4.3.3), individual reactions for each studied substrate concentration are monitored in a maximum of three different occasions (Figure 3.7A). Judging by the R^2 evaluation (Figure 3.7B), the experimental data is well described by the LM equation, whereas the overall score (Figure 3.7C) indicates that nothing can be concluded about the possible occurrence of assay interferences. Importantly, interferENZY can still produce accurate estimates of K_m^{app} and V^{app} using kinetic data collected after the initial period of constant reaction rate.

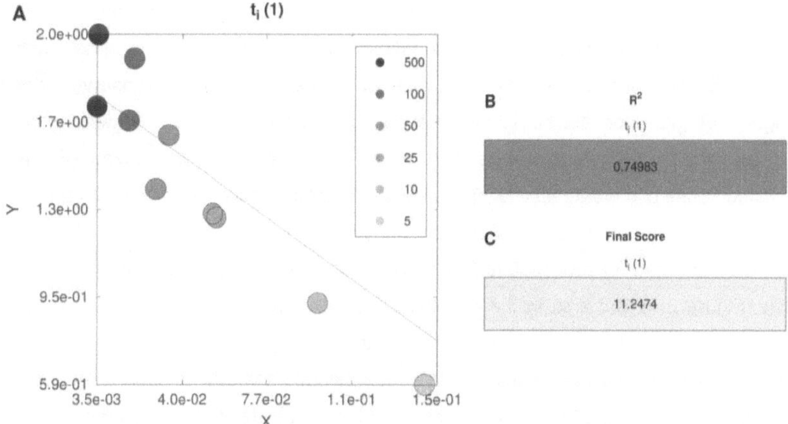

Figure 3.7. Example of interferENZY analysis of endpoint assays. (A) Linearized reaction coordinates of L-cysteine ($S_0 = [5 - 500]$ µM) catalysis by IscS ($E_0 = 1$ µM) in the presence of protein scaffold IscU (2 µM); see Chapter 4, Sections 4.2.1 and 4.3.3. Line: linear regression obtained using all LM coordinates. (B-C) Heatmap representations of (B) the coefficient of determination R^2 and (C) the interferENZY final score. The study of different t_i scenarios and of Kendall rank correlation coefficients was not performed since only 3 time points were available for each individual reaction. The determination of kinetic parameters requires no previous knowledge about what reaction phase is monitored.

3.7. Summary and outlook

Experimentalists willing to report their enzymology data may eventually face some of the following misgivings: were the initial reaction rates well measured? Was the period of constant rate missed? Were the experimental conditions fully stabilized? Are hidden interferences affecting my assay? Does the enzymatic reaction really follow a simple mechanism of the Briggs-Haldane type? What should be the choice: initial rate or full progress curve analysis? Which numerical methods should be adopted? How good is my enzymatic assay?

Casting away these doubts is now simpler with interferENZY, a public webserver into which crude experimental data can be uploaded, automatically tested for the presence of major assay interferences and, if the enzymatic assay is validated, numerically analyzed in order to estimate unbiased kinetic parameters. Different enzymatic systems were selected as real case-studies illustrating some of the more important features of the webserver. Since interferENZY analysis is not restricted to any particular stage of the reaction, it becomes possible to use data processing tools to remove signal noise without the risk of losing useful information (Figure 3.3). For the same reason, eventually missing the initial period of constant rate no longer represents an obstacle for kinetic parameter estimation in either continuous (Figure 3.5) and discontinuous (Figure 3.7) assays. The interferENZY webserver adopts the stringent LM criteria of assay validation that can identify and handle small interferences such as those arising from temperature fluctuations when reactants are mixed (Figure 3.5). Nonconformities of the dataset with single substrate, single active site mechanisms of enzyme kinetics visibly affect the statistical estimators used to compute the interferENZY final score. This was illustrated for procaspase-3, whose catalytic activity gradually declined during the assay (Figure 3.4), but also in the detection of subtle kinetic fingerprints associated to the isomerization of α-thrombin in the presence of an anticoagulant (Figure 3.6). Since the input data are subjected to a tight quality control before kinetic analysis, there is a reduced risk of fitted parameters being affected by unaccounted bias or by stealth interferences accumulated over the time-course. Besides the application examples hereby discussed, interferENZY can be used to quantitatively characterize the effect of novel inhibitors/activators of enzymes of interest, e.g., in drug discovery. In practical terms, the 17 basic enzyme modifier mechanisms described by the General Modifier Mechanism [27 (p.210), 31] are still well described by the LM equation (Eq. 3.2) provided that quasi-equilibrium conditions are attained and that the assay is free from extraneous interferences [1] (see Chapter 2, Section 2.4). Dissociation constants K_i can thus be estimated from the parameters K_m^{app} and V^{app} obtained for different concentrations of inhibitor.

3.8. Conclusions

The interferENZY webserver allows the validation and analysis of typical kinetic datasets obtained at different substrate concentrations and fixed enzyme concentration. The subjective component normally associated to initial rate analysis is eliminated by using automatically validated portions of measured progress curves. As opposed to blind curve-fitting procedures, interferENZY employs enzymology principles to analyze experimental data and detect unaccounted interferences ranging from instrumental drift to enzyme inactivation or inhibition. This publicly available and free platform was conceived as a useful tool for assay developers, as well as for enzymologists interested in reporting their kinetic results in a reproducible and accurate manner.

3.9. <u>Appendix</u>

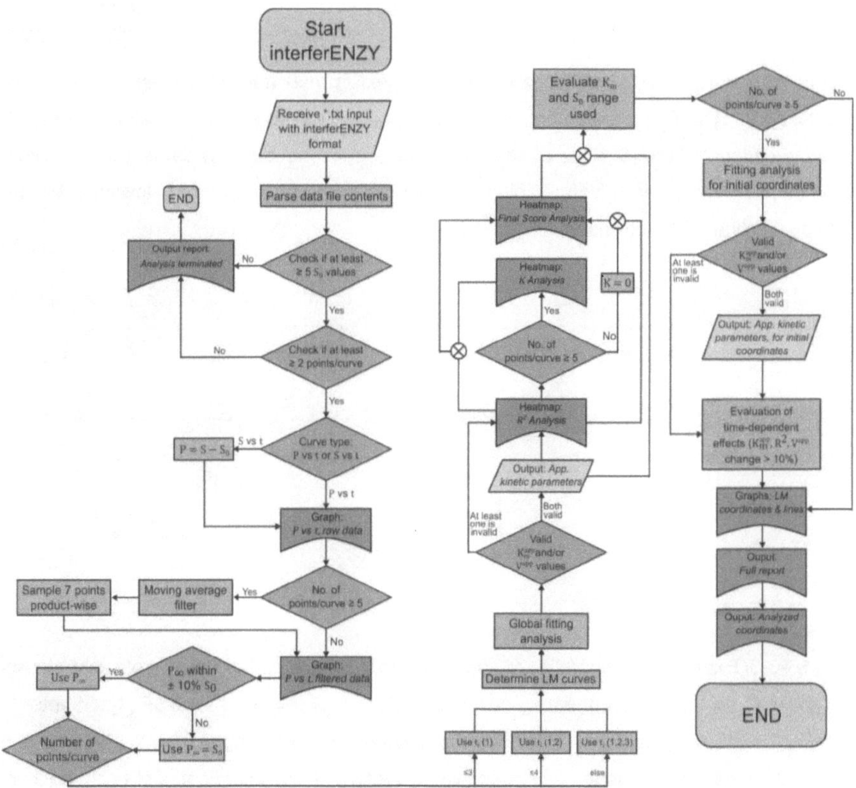

Figure A3.1. Flowchart depicting the analysis steps in the interferENZY script.

①	0.01		0.1		0.5		1		2.5		5	
②	0.01											
③	h											
④	uM											
⑤	Enzyme Assay											
⑥	0	0	0	0	0	0	0	0	0	0	0	0
	0.04	0.00	0.03	0.01	0.04	0.03	0.04	0.05	0.04	0.07	0.04	0.09
	0.08	0.00	0.06	0.01	0.07	0.06	0.07	0.10	0.07	0.14	0.08	0.18
	0.12	0.00	0.09	0.02	0.11	0.09	0.11	0.14	0.11	0.20	0.12	0.27
	0.16	0.00	0.12	0.03	0.14	0.12	0.14	0.19	0.14	0.27	0.16	0.34
	0.21	0.00	0.15	0.03	0.18	0.14	0.18	0.23	0.18	0.33	0.20	0.42
	Time	Measurements	Time	Measurements	Time	Measurements	Time	Measurements	Time	Measurements	Time	Measurements

Figure A3.2. Organization of the tab-separated (text format) dataset file used as input for interferENZY. The first 5 lines contain (1) tab-separated substrate concentration values, (2) enzyme concentration, (3) units of time, (4) units of concentration, (5) dataset name. Time-concentration curves are given in column pairs from line (6) downwards following the order indicated in line (1) for corresponding S_0 values.

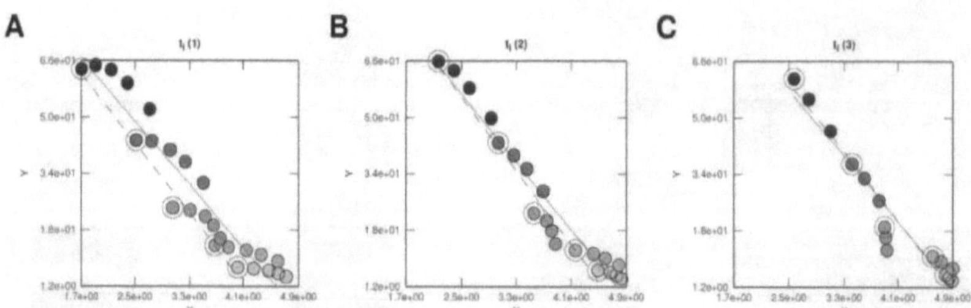

Figure A3.3. Example of automatic LM application to a dataset of the catalysis of substrate Ac-DEVD-AMC ($S_0 = [3.125, 50]$ μM) by caspase-3 ($E_0 = 1$ U) (Figure 3.5D-3.5F [1], Chapter 2, Section 2.5.3). (A-C) Symbols: linearized coordinates obtained using (A) $i = 1$, (B) $i = 2$, and (C) $i = 3$ as the LM first instant (t_i); open symbols refer to initial coordinates of each reaction. Lines: linear regressions obtained using (red) all or (blue) initial-only LM coordinates.

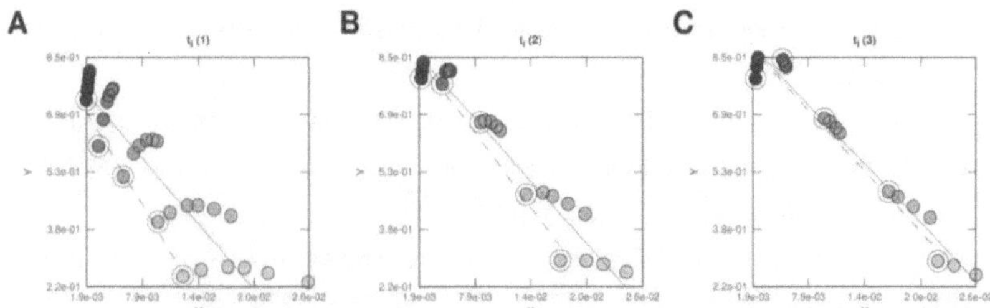

Figure A3.4. Example of automatic LM application to a dataset of the catalysis of substrate Tos-Gly-Pro-Arg-p-nitroanilide ($S_0 = 400, 200, 100, 50$ and 25 µM, color-coded as in Figure 3.6) by α-thrombin ($E_0 = 0.15$ nM [1,30], Chapter 2, Sections 2.5.4 and 2.9). (A-C) Symbols: linearized coordinates obtained using (A) $i = 1$, (B) $i = 2$, and (C) $i = 3$ as the LM first instant (t_i); open symbols refer to initial coordinates of each reaction. Lines: linear regressions obtained using (red) all or (blue) initial-only LM coordinates.

References

1. Pinto, M.F., *et al.* (2019), *A simple linearization method unveils hidden enzymatic assay interferences.* Biophysical Chemistry. 252:106193.

2. Swainston, N., *et al.* (2018), *STRENDA DB: enabling the validation and sharing of enzyme kinetics data.* FEBS Journal. 285(12):2193-2204.

3. Tipton, K.F., *et al.* (2014), *Standards for Reporting Enzyme Data: The STRENDA Consortium: What it aims to do and why it should be helpful.* Perspectives in Science. 1:131–137.

4. Zimmerle, C.T. and Frieden, C. (1989), *Analysis of progress curves by simulations generated by numerical integration.* The Biochemical journal. 258(2):381-387.

5. Barshop, B.A., Wrenn, R.F., and Frieden, C. (1983), *Analysis of numerical methods for computer simulation of kinetic processes: development of KINSIM--a flexible, portable system.* Analytical biochemistry. 130(1):134-145.

6. Kuzmič, P. (1996), *Program DYNAFIT for the Analysis of Enzyme Kinetic Data: Application to HIV Proteinase.* Analytical Biochemistry. 237(2):260-273.

7. Johnson, K.A., Simpson, Z.B., and Blom, T. (2009), *Global Kinetic Explorer: A new computer program for dynamic simulation and fitting of kinetic data.* Analytical Biochemistry. 387(1):20-29.

8.   Choi, B., Rempala, G.A., and Kim, J.K. (2017), *Beyond the Michaelis-Menten equation: Accurate and efficient estimation of enzyme kinetic parameters.* Scientific Reports. 7(1):17018.

9.   Bäuerle, F., Zotter, A., and Schreiber, G. (2016), *Direct determination of enzyme kinetic parameters from single reactions using a new progress curve analysis tool.* Protein Engineering, Design and Selection. 30(3):151-158.

10.   Bevc, S., *et al.* (2011), *ENZO: a web tool for derivation and evaluation of kinetic models of enzyme catalyzed reactions.* PloS one. 6(7):e22265-e22265.

11.   Olp, M.D., Kalous, K.S., and Smith, B.C. (2019), *An online tool for calculating initial rates from continuous enzyme kinetic traces.* bioRxiv:700138.

12.   Eicher, J., Snoep, J., and Rohwer, J. (2012), *Determining Enzyme Kinetics for Systems Biology with Nuclear Magnetic Resonance Spectroscopy.* Metabolites. 2(4):818.

13.   Duggleby, R.G. (2001), *Quantitative analysis of the time courses of enzyme-catalyzed reactions.* Methods. 24(2):168-174.

14.   Orsi, B.A. and Tipton, K.F. (1979), *Kinetic analysis of progress curves.* Methods in Enzymology. 63:159-83.

15.   Cornish-Bowden, A., (2012), *Fundamentals of Enzyme Kinetics*, 4th ed. Wiley-Blackwell (Weinheim, Germany), pp.87.

16.   Pinto, M.F., *et al.* (2015), *Enzyme kinetics: the whole picture reveals hidden meanings.* FEBS Journal. 282(12):2309-2316.

17.   Schnell, S. and Mendoza, C. (1997), *Closed Form Solution for Time-dependent Enzyme Kinetics.* Journal of Theoretical Biology. 187(2):207-212.

18.   Ferreira, C., *et al.* (2019), *Protein crystals as a key for deciphering macromolecular crowding effects on biological reactions.* Submitted.

19.   Hanson, S.M. and Schnell, S. (2008), *Reactant Stationary Approximation in Enzyme Kinetics.* The Journal of Physical Chemistry A. 112:8654-8658.

20.   Turrens, J.F. and Boveris, A. (1980), *Generation of superoxide anion by the NADH dehydrogenase of bovine heart mitochondria.* Biochemical Journal. 191(2):421.

21.   Markert, C.L. (1984), *Lactate dehydrogenase. Biochemistry and function of lactate dehydrogenase.* Cell Biochemistry and Function. 2(3):131-134.

22.   McComb, R.B., *et al.* (1976), *Determination of the molar absorptivity of NADH.* Clinical Chemistry. 22(2):141-50.

23.   Bisswanger, H. (2014), *Enzyme assays.* Perspectives in Science. 1(1–6):41-55.

24.   Chaplin, M.F. and Bucke, C., (1990), *Enzyme technology.* CUP Archive, pp.32.

25.   Reed, M.C., Lieb, A., and Nijhout, H.F. (2010), *The biological significance of substrate inhibition: a mechanism with diverse functions.* Bioessays. 32(5):422-9.

26.   Roskoski, R., (2014), Enzyme Assays☆. Elsevier, Chp.

27. Baici, A., (2015), *Kinetics of Enzyme-Modifier Interactions*, 1st ed. Springer.

28. Kendall, M.G. (1938), *A new measure of rank correlation.* Biometrika. 30(1-2):81-93.

29. van Doorn, J., *et al.* (2018), *Bayesian Inference for Kendall's Rank Correlation Coefficient.* The American Statistician. 72(4):303-308.

30. Watson, E.E., *et al.* (2019), *Rapid assembly and profiling of an anticoagulant sulfoprotein library.* Proceedings of the National Academy of Sciences. 116(28):13873.

31. Botts, J. and Morales, M. (1953), *Analytical description of the effects of modifiers and of enzyme multivalency upon the steady state catalyzed reaction rate.* Transactions of the Faraday Society. 49(0):696-707.

Chapter 4.

Effect of bacterial frataxin CyaY on the Enzyme Kinetics of IscS:IscU

4.2. <u>Introduction</u>

4.2.1. Bacterial frataxin CyaY & Fe-S cluster biogenesis machinery

FRDA is a neurodegenerative disease caused by deficiency of a small essential protein that is highly conserved from bacteria to humans – frataxin. This deficiency is caused by an abnormal expansion of a non-coding GAA triplet repeat in the first intron of the *FRDA* gene. This leads to lower expression levels of frataxin, due to the heterochromatization of the corresponding locus [1,2]. The function of frataxin is generally associated to iron metabolism: its deficiency results in the accumulation of mitochondrial iron and a loss of activity in Fe-S cluster dependent proteins [3,4]. Frataxin binds to ferrochelatase [5-8] and to vital components of the Fe-S cluster assembly machinery [9-11], thus being implicated in heme metabolism and Fe-S cluster formation.

Fe-S clusters are evolutionary ancient and versatile inorganic cofactors able to act as catalysts or redox sensors in a large variety of cell pathways. Fe-S cluster biogenesis is initiated by a pyridoxal phosphate (PLP)-dependent cysteine desulfurase, which produces sulfur (S^0) from L-cysteine (also producing L-alanine) and is then transferred as sulfide (S^{2-}) to a protein scaffold. Iron (Fe^{2+}) and S^{2-} come together on this scaffold to give rise to Fe-S clusters [12,13] (Figure 4.1). While the source of iron in this process is not known with certainty, frataxin has been suggested to act as an iron chaperone [12]. However, recent work strongly indicates that, more than an iron chaperone, frataxin actively participates in Fe-S cluster biogenesis. For example, the bacterial ortholog of frataxin (CyaY) was shown to act as a molecular regulator inhibiting the production of Fe-S clusters by the corresponding *Escherichia coli* protein machinery. [14]. In this organism, the genes encoding the protein components of the Fe-S cluster machinery are grouped in the *isc* operon, except for CyaY. The *isc* operon is controlled by IscR, a Fe-S cluster binding transcriptional repressor. This operon includes the genes encoding for the cysteine desulfurase IscS, the protein scaffold IscU, as well as IscA, HscB and HscA (molecular chaperones), Fdx (ferredoxin) and YfhJ (modulator also known as IscX) [15].

In the present work, the bacterial Fe-S cluster biogenesis machinery was used as a model to study the enzymatic processes associated with Fe-S cluster production. More specifically, the kinetics of sulfur production by IscS (EC 2.8.1.7) in the presence of IscU and CyaY are analyzed in light of the general modifier mechanism (GMM) [16] for qualitative and quantitative characterization of CyaY's inhibitory effect.

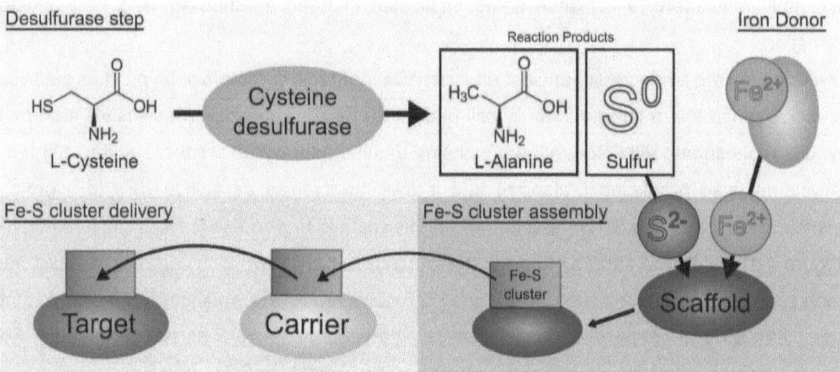

Figure 4.1. General process of Fe-S cluster biogenesis. A cysteine desulfurase (e.g. IscS) catalyzes the conversion of L-cysteine to L-alanine, obtaining a sulfur group (S^0) which is transferred in the form of sulfide (S^{2-}) to a scaffold protein (e.g. IscU). An iron donor provides the iron (Fe^{2+}) for Fe-S cluster assembly. The cluster is then transferred from the scaffold to a carrier protein (e.g. HscB/HscA) which transports and delivers it to its final apo-targets. Adapted from Roche *et al.* (2013) [17].

4.2.2. General Modifier Mechanism (GMM)

With the goal of analyzing the effect of modulator compounds (e.g. inhibitors, activators) on enzyme activity, Botts and Morales (1953) [16] developed a kinetic model later designated as the General Modifier Mechanism (Scheme 4.1). It admits the possible formation of two substrate (S)-bound intermediates, namely the enzyme-substrate ES complex, and the ternary complex between the enzyme E, S and modifier compounds X (ESX).

A

$$E + S \underset{k_{-1}}{\overset{k_1}{\rightleftharpoons}} ES \overset{k_2}{\rightarrow} E + P$$
$$+ \qquad\qquad +$$
$$X \qquad\qquad X$$
$$k_{-3}\updownarrow k_3 \qquad k_{-4}\updownarrow k_4$$
$$EX + S \underset{k_{-5}}{\overset{k_5}{\rightleftharpoons}} ESX \overset{k_6}{\rightarrow} EX + P$$

B

$$E + S \overset{K_S}{\rightleftharpoons} ES \overset{k_2}{\rightarrow} E + P$$
$$+ \qquad\qquad +$$
$$X \qquad\qquad X$$
$$K_{Sp}\updownarrow \qquad K_{Ca}\updownarrow$$
$$EX + S \overset{\alpha K_S}{\rightleftharpoons} ESX \overset{\beta k_2}{\rightarrow} EX + P$$

Scheme 4.1. Different representations of the GMM. Notation (A) adopts rate constants for all reaction steps, while notation (B) uses dissociation constants and distinguishes the specific and catalytic components of modifier activity. The dissociation constants are defined as $K_s = k_{-1}/k_1$ (substrate dissociation constant), $K_3 = k_{-3}/k_3 = K_{Sp} = K_X$ (dissociation constant of the specific component), $K_4 = k_{-4}/k_4 = K_{Ca} = \alpha K_{Sp}$ (dissociation constant of the catalytic component), $K_5 = k_{-5}/k_5 = \alpha K_S$. Parameters $\alpha = K_{Ca}/K_{Sp}$ and β are, respectively, the reciprocal allosteric coupling constant between modifier and substrate, and the factor by which the modifying compound alters the catalytic constant k_2. The reverse reactions k_{-2} and k_{-6} are not shown since only the initial rate is considered ($[P] = 0$) [18, p.70].

From this model, the authors derived an exact rate equation under the steady-state assumption. Due to the presence of the two substrate-bound intermediate species (ES and ESX), the steady-state rate equation in question contains squared terms of concentration, which can be simplified by applying the following practical assumptions: (i) quasi-equilibrium conditions for all steps, (ii) generalized microscopic reversibility [16,19], and (iii) quasi-equilibrium assumption for the binding of modifier [18, pp.76-77] Under these conditions, the rate equation for the GMM simplifies to (Eq. 4.1):

$$v = \frac{k_2\left(1 + \beta \frac{[X]}{\alpha K_X}\right)[E]_t[S]}{K_m\left(1 + \frac{[X]}{K_X}\right) + [S]\left(1 + \frac{[X]}{\alpha K_X}\right)} \tag{4.1}$$

where $[E]_t$, $[S]$, and $[X]$ are the concentration values of total enzyme, initial substrate, and present modifier, K_m is the Michaelis constant, and K_X and αK_X are the dissociation constants for the specific and catalytic portions of the GMM [18, p.131].

Eq. 4.1 is conveniently rewritten as a Michaelis-Menten (MM)-like equation using the following apparent MM parameters (Eqs. 4.2a-4.2c) [18, pp.130-131]:

$$k_{cat}^{app} = k_2 \times \frac{1 + \beta \frac{[X]}{\alpha K_X}}{1 + \frac{[X]}{\alpha K_X}} \qquad (4.2a)$$

$$K_m^{app} = K_m \times \frac{1 + \frac{[X]}{K_X}}{1 + \frac{[X]}{\alpha K_X}} \qquad (4.2b)$$

$$\left(\frac{k_{cat}}{K_m}\right)^{app} = \frac{k_2}{K_m} \times \frac{1 + \beta \frac{[X]}{\alpha K_X}}{1 + \frac{[X]}{K_X}} \qquad (4.2c)$$

4.2.3. Classification of enzyme-modifier interactions

Allosteric and non-allosteric interactions between enzyme, substrate and modulator can be analyzed in the light of the GMM [18, p.65]. As systematized by Baici (2015) [18], there are 17 types of basic mechanisms of inhibition and nonessential activation by which these interactions occur (Table 4.1). This simple but comprehensive nomenclature replaces the commonly used terms "competitive" and "uncompetitive" by "specific" and "catalytic", in accordance with a previous study by Cárdenas and Cornish-Bowden (1989) [20]. The terms "balanced" and "dual" are also introduced to refer to mechanisms where the specific and catalytic components have equal weights, and mechanisms that may interchange from inhibition to activation (or viceversa) depending on the critical substrate concentration for which the reaction rate is the same in the presence or absence of modifier [18, p.161], respectively [18, p.134]. This classification primarily attends to the value of α and then to the value of β. The value of α allows the distinction between the predominantly specific ($1 < \alpha < \infty$), catalytic ($0 < \alpha < 1$), and balanced ($\alpha = 1$) character of a modifier. The value of β, together with its relationship with α, determines whether a given mechanism is of inhibition or activation. Finally, modifying mechanisms can be described as linear or hyperbolic depending on the observed tendency for the plots of $1/k_{cat}^{app}$ and $1/(k_{cat}/K_m)^{app}$ versus $[X]$ (Eqs. 4.3a-4.3c) [18, p.131]:

$$\frac{1}{k_{cat}^{app}} = \frac{1}{k_2} \times \frac{1 + \frac{[X]}{\alpha K_X}}{1 + \beta \frac{[X]}{\alpha K_X}} \qquad (4.3a)$$

$$\frac{1}{(k_{cat}/K_m)^{app}} = \frac{K_m}{k_2} \times \frac{1 + \dfrac{[X]}{K_X}}{1 + \beta\dfrac{[X]}{\alpha K_X}} \tag{4.3b}$$

A modifier mechanism is considered linear or hyperbolic when at least one of the dependencies of $1/k_{cat}^{app}$ and $1/(k_{cat}/K_m)^{app}$ on $[X]$ has a linear or hyperbola shape [18, p. 131].

Table 4.1. Basic enzyme-modifier interactions of inhibition and nonessential activation. A total of 17 basic mechanisms are presented along with the respective values of α and β. Adapted from [18, p.210].

No.	Mechanism	Acronym	α	β	Other
1	Linear specific inhibition	LSpI	∞[a]	0	$K_{Ca} = \infty$
2	Linear catalytic inhibition	LCaI	0[b]	0	$K_{Sp} = \infty$
3	Linear mixed, predominantly specific inhibition	LMx(Sp > Ca)I	$(1, \infty)$	0	
4	Linear mixed, predominantly catalytic inhibition	LMx(Sp < Ca)I	$(0,1)$	0	
5	Linear mixed, balanced inhibition	LMx(Sp = Ca)I	1	0	
6	Hyperbolic specific inhibition	HSpI	$(1, \infty)$	1	
7	Hyperbolic mixed, predominantly specific inhibition	HMx(Sp > Ca)I	$(1, \infty)$	$(0,1)$	
8	Hyperbolic catalytic inhibition	HCaI	$(0,1)$	$(0,1)$	$\alpha = \beta$
9	Hyperbolic mixed, predominantly catalytic inhibition	HMx(Sp < Ca)I	$(0,1)$	$(0,1)$	$\alpha > \beta$
10	Hyperbolic mixed, balanced inhibition	HMx(Sp = Ca)I	1	$(0,1)$	
11	Hyperbolic mixed, dual modification (inhibition $\rightarrow$ activation)	HMxD(I/A)	$(1, \infty)$	$(1, \infty)$	$\alpha > \beta$
12	Hyperbolic catalytic activation	HCaA	$(1, \infty)$	$(1, \infty)$	
13	Hyperbolic mixed, predominantly specific activation	HMx(Sp > Ca)A	$(1, \infty)$	$(1, \infty)$	$\alpha < \beta$
14	Hyperbolic mixed, dual modification (activation $\rightarrow$ inhibition)	HMxD(A/I)	$(0,1)$	$(0,1)$	$\alpha < \beta$
15	Hyperbolic specific activation	HSpA	$(0,1)$	1	
16	Hyperbolic mixed, predominantly catalytic activation	HMx(Sp < Ca)A	$(0,1)$	> 1	
17	Hyperbolic mixed, balanced activation	HMx(Sp = Ca)A	1	> 1	

[a]Mechanism is better defined considering $K_{Ca} = \infty$.
[b]Mechanism is better defined considering $\alpha \rightarrow 0$ and $K_{Sp} \rightarrow \infty$

Kinetic mechanisms that may otherwise be vaguely classified as "mixed" [18, p.129] or as a generic assortment of other mechanisms [18, p.133] are unambiguously characterized both qualitatively and quantitatively by the GMM. This is next illustrated for the analysis of kinetic modulation effects of CyaY on the desulfurase step of Fe-S cluster biogenesis.

4.3. <u>Methods</u>

4.3.1. Protein Production

IscS, IscU and CyaY constructs previously subcloned into pET-derived plasmid vectors (performed by Annalisa Pastore Group) were used as fusion proteins containing a C-terminal His$_6$-tag or His$_6$-tagged glutathione-S-transferase (GST) tag, for IscS and IscU/CyaY, correspondingly. All constructs contained a tobacco etch virus (TEV) protease cleavage site for tag removal. IscS, IscU and CyaY were overexpressed in *E. coli* BL21(DE3) as previously described [14,21,22], with minor alterations. Bacteria expressing IscU were grown in LB medium containing 8.3 µM ZnSO$_4$ to stabilize its fold [23-25]. For protein expression, cells were induced for 3-4 hours at 37 °C by the addition of 0.5 mM isopropyl b-D-thiogalactopyranoside (IPTG) after the cultures reached an optical density (OD) of 0.6-0.8 at 600 nm, in the presence of kanamycin (50 µg/mL). Cell pellets were harvested, resuspended in lysis buffer (20 mM Tris pH 8, 150 mM NaCl, 10 mM imidazole, 0.5% v/v igepal, 0.5 mM TCEP or 5 mM β-mercaptoethanol) containing lysozyme, DNaseI and protease inhibitors, and frozen at -80 °C. After thawing, the suspension was sonicated and centrifuged to recover the soluble fraction. Following protein purification steps were carried out in the presence of 0.5 mM TCEP or 5 mM β-mercaptoethanol. Overexpressed proteins were purified by affinity-chromatography using nickel-nitrilotriacetic acid (Ni-NTA) agarose gel (Qiagen or Agarose Bead Technologies), and cleaved from the corresponding tags by TEV protease (in-house) under dialysis overnight at 4 °C. The mixture was then passed through the Ni-NTA agarose gel and further purified using 16/60 Superdex G75 or 16/60 HiPrep S-100 columns (GE Healthcare). Samples were eluted in a solution of 50 mM Tris-HCl buffer (pH 7.5), containing 150 mM NaCl and 0.5 mM TCEP or 5 mM β-mercaptoethanol. Purity of all proteins was checked using SDS-PAGE after each purification step and for the final product. Final protein stock concentrations were determined spectrophotometrically using the molar absorption coefficients ε_{280}(IscS) = 41370 M^{-1}cm^{-1}, ε_{280}(IscS) = 11460 M^{-1}cm^{-1}, ε_{280}(IscS) = 28990 M^{-1}cm^{-1}, obtained by ProtParam analysis (ExPASy) of protein sequences without fusion tags.

4.3.2. Alanine quantification by mass spectrometry

Enzymatic reactions were performed in Tris 50 mM pH 7.5, 150 mM NaCl, with 1 µM IscS, 2 µM IscU, 0,10 µM CyaY, 3 mM DTT. Reactions were started by addition of substrate L-cysteine (Sigma) and quenched after different time intervals at 37 °C in a stopped assay procedure. Reaction quenching was done by removing an aliquot of the ongoing reaction (full reaction volume of 800 µL) and adding it to acetonitrile/formic acid (100:0.1) in a 1:4 ratio. The samples were then cooled on ice. To each sample an internal standard consisting of a known amount of isotope labelled alanine (L-alanine, 2,3-$^{13}C_2$, 99%, Cambridge Isotope Laboratories) was added. Samples were then centrifuged at 17,000xg for 5 min at 4 °C. The supernatants were collected and dried using a Savant SpeedVac Concentrator (Thermo Scientific, United Kingdom) at 45 °C, 0.01 Torr, for approximately 2 hours. Alanine quantification was performed at Professor Luigi Servillo's lab in Università della Campania, Naples, Italy. Samples were dissolved in 100 µL of 0.1% HCOOH in water, and injected (10 µL) on the column for high-performance liquid chromatography/electrospray ionization tandem mass spectrometry (HPLC-ESI/MS). Peak areas for detection of labelled and unlabeled alanine were obtained and alanine was quantified using the known internal standard as reference. Assays were performed in triplicate.

4.3.3. Sulfide quantification by methylene blue assay

A modified version of the method described by Siegel (1965) [26] was employed to quantify S^{2-} as a product of the enzymatic reaction, through its conversion to hydrogen sulfide (H_2S) gas and subsequent incorporation into methylene blue. Enzymatic reactions were performed in Tris 50 mM pH 7.5, 150 mM NaCl, with 1 µM IscS, 2 µM IscU, 0-10 µM CyaY, 3 mM DTT. Reactions were started by addition of substrate L-cysteine (Sigma) and quenched at different time intervals at 37 °C or 20 °C in a stopped assay procedure. Replicate samples with reaction volume 800 µL were analyzed for each intended timepoint. All solutions were equilibrated to reaction temperature for at least 30 minutes prior usage. Reaction quenching was done by the addition of 100 µL of 20 mM N,N-Dimethyl-p-phenylenediamine dihydrochloride (DPD·2HCl) in 7.2 N HCl followed by the addition of 100 µL of 30 mM $FeCl_3$ in 1.2 N HCl. The samples were manually shaken for 1 minute and incubated in the dark at 20 °C for at least 1 hour for methylene blue development. H_2S gas loss was avoided by using 10 mm wide rubber septa stoppers in 10x75 mm test tubes for each sample: rubber stoppers were pierced using Hamilton® GASTIGHT® syringes (1710 RN, 100:1 µl, ga22s/51mm/pst2) to add the reagents for reaction quenching. After methylene blue development, samples were transferred to new tubes and centrifuged for 5 minutes at >17,000xg. Methylene blue was then quantified by absorbance at λ=670 nm [27] using a Cary 50 Bio (Varian) or Shimadzu UV-1800 spectrophotometer. Assays were performed in duplicate or triplicate. A calibration curve was built using solutions with known concentration of Na_2S in Tris 50 mM pH 7.5, 150 mM NaCl (without DTT) for conversion of the resulting methylene blue absorbance to concentration of product S^{2-}.

4.3.4. Data Analysis

A custom script was built using Mathworks® MATLAB R2018a (Natick, MA, USA) for data analysis. Initial reaction rate v_0 values were determined by linear regression of initial and intermediate reaction timepoints for the studied conditions, using the *fitlm* function. Apparent parameter K_m^{app}, k_{cat}^{app} and k_{cat}^{app}/K_m^{app} determination was performed by nonlinear fitting of Eq. 4.4 using the *fitnlm* function to each determined v_0 vs S_0 curve.

$$v_0 = \frac{k_{cat}^{app} E_0 S_0}{K_m^{app} + S_0} \tag{4.4}$$

A nonlinear least-squares method with the function *lsqcurvefit* was employed for the numerical fittings of (i) Eqs. 4.2a-4.2c to apparent parameter versus modifier concentration dependencies, (ii) Eq. 4.5 (see Section 4.4.3) to v_0/v_X ratios and (iii) Eq. 4.1 to the full dataset of v_0 vs S_0 curves. Numerical procedures using *lsqcurvefit* can only be performed using velocity data obtained for the same set of substrate and CyaY concentration values. For this reason, the v_0 value obtained for [CyaY] = 0 µM and [L-cys] = 75 µM could not be used for global fitting (iii). Standard errors for 95% confidence interval for parameters obtained from nonlinear fitting were determined using the *nlparci* function.

4.4. <u>Results and Discussion</u>

4.4.1. Assay optimization

The methylene blue assay successfully detected the amount of S^{2-} produced from L-cysteine by IscS, in the presence of IscU and different concentrations of CyaY (0, 0.5, 1, 5 and 10 µM) over different timepoints (Figure 4.2A). A concentration-dependent inhibitory effect of CyaY over the kinetic behavior of IscS in the presence of IscU is observed. Relatively to previous similar results in the context of the overall mechanism of Fe-S cluster formation [14], our results indicate that the initial desulfurase step is certainly inhibited by CyaY while it remains to be studied what is the role of CyaY in the subsequent step of Fe-S cluster assembly.

Before arriving to these conclusions, the following assay optimization steps were taken:

(i) Calibration curves firstly obtained using Na_2S or Li_2S for S^{2-} quantification failed to adequately replicate the conditions in which the enzymatic assays were performed. This was because methylene blue is colored blue in an oxidizing environment and becomes clear/colorless if exposed to reducing agents [28]. In fact, these test calibration curves were obtained in a buffer with the same composition as that of the enzymatic reactions containing 3 mM DTT. The addition of DTT is not expected to have any deleterious effect in the development of methylene blue in the enzymatic reaction environment due to the quick expenditure of this reducing agent by the reaction components; however, for the determination of the calibration curve, only Na_2S or Li_2S were present in the reaction buffer together with DTT. Subsequent determination of calibration curves in the absence of DTT recovered the expected behavior for S^{2-} quantification.

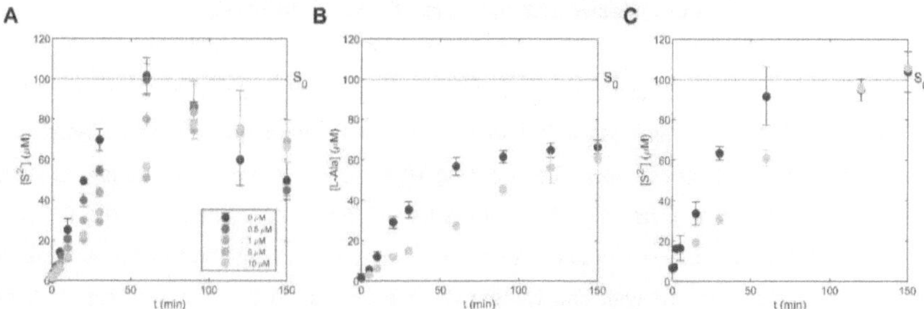

Figure 4.2. Inhibitory effect of CyaY on the desulfurase activity of IscS in the presence of IscU. (A-C) Reaction progress curves for the catalysis of [L-cysteine] = 100 µM to S^{2-} or L-alanine for [IscS] = 1 µM, [IscU] = 2 µM, and [CyaY] = 0 − 10 µM measured using: (A) T = 37 °C, S^{2-} quantification by the methylene blue assay with a preliminary calibration curve obtained by measurement of methylene blue for reactions with known [L-cys] and [CyaY] = 0 µM; (B) T = 37 °C, alanine quantification by HPLC-ESI/MS using isotope-labelled alanine as internal standard; (C) T = 20 °C, S^{2-} quantification by the methylene blue assay with Na_2S calibration curve. Symbols and bars indicate means and standard deviations for (A,B) three and (C) two replicates. Dotted lines indicate the initial concentration of substrate ($P_\infty = S_0$).

(ii) Reaction conditions had to be additionally optimized due to potential enzyme inactivation and/or deterioration of reaction environment conditions that are associated to the product concentration decay observed over the second half of the progress curves in Figure 4.2A. To overcome the limitations of the methylene blue assay, alanine quantification was tested in view of the higher stability of alanine relatively to the labile product S^{2-}. An enzymatic reporter system was tried involving the conversion of alanine to pyruvate by alanine transferase, with subsequent catalysis of pyruvate to several products including H_2O_2, and a final step involving the reaction of H_2O_2 with a fluorogenic probe [29]. This method was not successfully optimized because the fluorogenic probe was inadvertently activated in the presence of reducing agents such as DTT or TCEP. These elements cannot be removed from the reaction environment as they ensure that the –SH functional groups present in catalytic cysteine aminoacids in IscS [30,31] and the –SH group in the substrate L-cysteine remain in the reduced form.

(iii) Mass spectrometry was then tested for alanine quantification (Figure 4.2B), although the final product concentration values obtained were again lower than the initial substrate concentration. This result again suggests the possible occurrence of phenomena like progressive enzyme inactivation or substrate loss.

(iv) The temperature of the reaction of 37 °C was lowered to 20 °C in subsequent time-course measurements as it is known that Tris buffer pH lowers considerably with temperature increase, and the reducing agent DTT decays more rapidly at higher temperatures (e.g. half-life of 1.4 hours for pH 8.5 at 20 °C decreases to 0.2 hours when temperature is increased to 40 °C [32]). After this change of procedures, the enzymatic reactions with and without CyaY ended up in the same expected plateau (Figure 4.2C). The reduction in temperature, while slightly decreasing reaction rates, advantageously reduces enzyme inactivation, product decay, and substrate loss over time.

4.4.2. Analysis of apparent kinetic parameters

The influence of substrate concentration on the initial reaction rates v_0 was determined for different concentrations of CyaY (Figure 4.3). Then, the effect of [CyaY] on the fitted parameters K_m^{app}, k_{cat}^{app} and k_{cat}^{app}/K_m^{app} (Eqs. 4.2a-4.2c) was investigated (Figure 4.4).

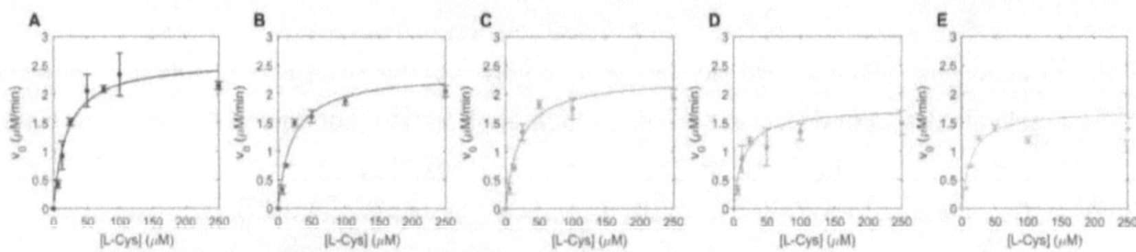

Figure 4.3. Initial reaction velocity versus substrate concentration curves (v_0 vs S_0) obtained for the catalysis of substrate [L-cysteine] (6.25, 12.5, 25, 50, 100, 250 μM) at 20 °C by [IscS] = 1 μM, in the presence of [IscU] = 2 μM, and [CyaY] values of (A) 0 μM, (B) 0.5 μM, (C) 1 μM, (D) 5 μM, and (E) 10 μM. Symbols and error bars: means and standard deviations of v_0 values obtained for two replicates; solid line: fit of the MM-like equation (Eq. 4.4) to experimental data.

Each of the 17 basic kinetic mechanisms defined by the GMM assumes a particular profile in terms of apparent parameter dependence on the modifier concentration, thus allowing a preliminary classification of the kinetic mechanism in question (Table 4.1). In the present case, hyperbolic descending trends are observed for the apparent parameters K_m^{app} and k_{cat}^{app} (and hyperbolic increasing trend for $1/k_{cat}^{app}$), while k_{cat}^{app}/K_m^{app} and K_m^{app}/k_{cat}^{app} exhibit constant lines. According to the GMM scenarios described in [18, p.136,138], this behavior is in accordance with a hyperbolic catalytic mechanism of inhibition. A direct interpretation of this type of mechanism in the light of Scheme 4.1 suggests that the binding of the catalytic inhibitor is preceded by the formation of the enzyme-substrate complex.

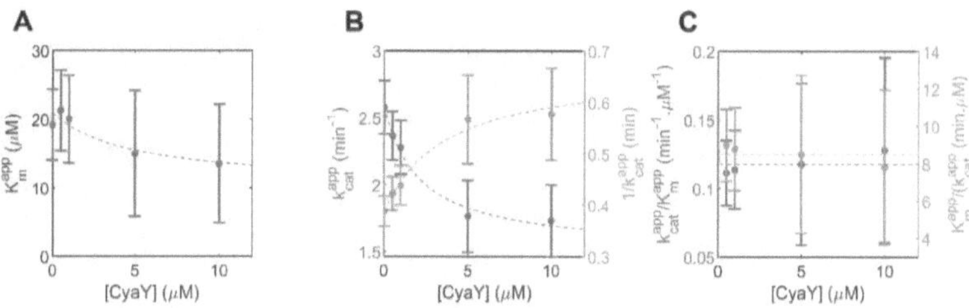

Figure 4.4. Dependence of apparent kinetic parameters (A) K_m^{app}, (B) k_{cat}^{app} (and reciprocal $1/k_{cat}^{app}$) and (C) k_{cat}^{app}/K_m^{app} (and reciprocal K_m^{app}/k_{cat}^{app}) on modifier concentration [CyaY] values of 0, 0.5, 1, 5 and 10 μM. Symbols and error bars: fitted apparent parameters and standard errors for 95% confidence level (from Figure 4.3). Dashed lines: fit of Eqs. 4.2a-4.2c and representation of Eqs. 4.3a-4.3b from fitted parameters.

4.4.3. Specific velocity plots

To further investigate the classification of the kinetic mechanism, the specific velocity plot method was employed (Figure 4.5). The primary specific velocity plot (Figure 4.5A) is based on the following linearized equation (Eq. 4.5), which results from algebraic manipulation of Eq. 4.1 and of the MM equation in the absence of modifier [33]:

$$\frac{v_0}{v_X} = \frac{[X]\left(\frac{1}{\alpha K_X} - \frac{1}{K_X}\right)}{1 + \beta \frac{[X]}{\alpha K_X}} \frac{\sigma}{1+\sigma} + \frac{1 + \frac{[X]}{K_X}}{1 + \beta \frac{[X]}{\alpha K_X}} \tag{4.5}$$

where $V = k_2 E_0$, $\sigma = S_0/K_m$, and $\sigma/(1+\sigma)$ is designated as specific velocity. When v_0/v_X is plotted against the specific velocity the result is a set of straight lines whose collective behavior can be assigned to its characteristic basic mechanism [18, p.156]. This graphical method is particularly useful in the identification of hyperbolic modifiers, as it allows the immediate appreciation of the linear/hyperbolic characters of the modifier mechanism [33]. In the present case, there is a large dispersion of velocity ratios (Figure 4.5A), which results from the amplification of the experimental error in v_0 determination in both independent and dependent variables of Eq. 4.5 [18, p. 158]. Moreover, the values of initial rates are less accurately estimated under conditions of low substrate concentrations for which the initial linear phase is rapidly elapsed. The positive slopes obtained in the primary plot confirm the catalytic character of CyaY's inhibitiory effect.

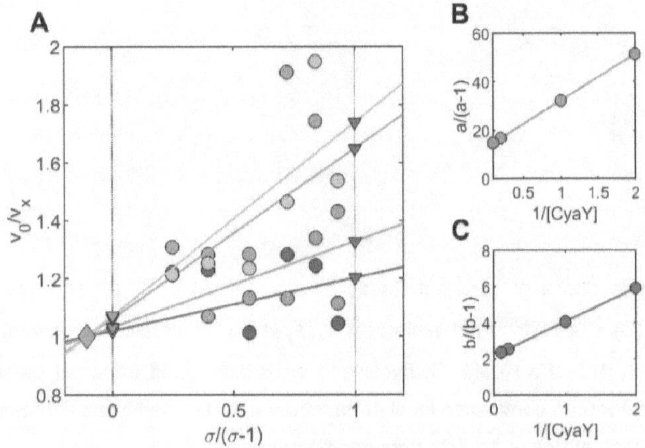

Figure 4.5. Specific velocity analysis. (A) Primary plot representation of the ratio between initial velocity in the absence and presence of modifier (v_0/v_X) as a function of the specific velocity $\sigma/(\sigma - 1)$. Circles: Measured velocity ratios for CyaY concentrations of 10, 5, 1, and 0.5 μM (color-coded as in Figure 4.3). Solid lines: nonlinear fitting of Eq. 4.5 to v_0/v_X vs $\sigma/(\sigma - 1)$ data. Orange and blue triangles: v_0/v_X intercepts for $\sigma/(\sigma - 1)$ values of 0 and 1, respectively; the intercepts correspond to the values of a and b. Green diamond: intersection between straight lines for $v_0/v_X = 1$. (B,C) Symbols: secondary plot representation of estimated values of (B) a (Eq. 4.6a), and (C) b (Eq. 4.6b) as functions of the reciprocal of [CyaY]. Solid lines: linear regressions.

As a step for the quantification of preliminary α, β and K_X parameters, extrapolated values of a and b were represented as a function of $1/[X]$ (Figure 4.5B and Figure 4.5C). The so-called secondary plots derive from the following linearized forms of Eq. 4.5 (Eqs. 4.6a,4.6b):

$$\frac{a}{a-1} = \frac{\alpha K_X}{\alpha - \beta}\frac{1}{[X]} + \frac{\alpha}{\alpha - \beta} \tag{4.6a}$$

$$\frac{\beta}{\beta-1} = \frac{\alpha K_X}{1-\beta}\frac{1}{[X]} + \frac{1}{1-\beta} \tag{4.6b}$$

The fitted values of $\alpha = 0.583$, $\beta = 0.5385$ and $K_X = 0.5385$ µM, and the behaviors observed for the $a/(a-1)$ vs. $1/[X]$ and $b/(b-1)$ vs. $1/[X]$ further point out to a mixed, predominantly catalytic mechanism of CyaY inhibition [18, p.158,160].

4.4.4. Final kinetic analysis

In the concluding part of the GMM analysis, Eq. 4.1 was fitted to the full set of v_0 vs S_0 curves using the preliminary values of α, β and K_X as initial estimations (Figure 4.6A). The newly fitted parameters of $\alpha = 0.671 \pm 0.183$, $\beta = 0.538 \pm 0.054$ are in line with the previous analysis; the partially mixed character of the inhibition mechanism is also confirmed by the K_m^{app} and k_{cat}^{app} plots (Figure 4.6B,4.6C), as well as by the hyperbolic descending and ascending trends now observed for k_{cat}^{app}/K_m^{app} and K_m^{app}/k_{cat}^{app}, respectively (Figure 4.6D). The determined values of $K_X = K_{Sp} = 1.84 \pm 0.89$ µM and $K_{Ca} = 1.23 \pm 0.69$ µM are characteristic of predominantly catalytic inhibitors ($K_{Ca} < K_{Sp}$).

IscS exists as a homodimer in solution [34], where each subunit of the dimer can bind one unit of IscU and/or CyaY [35]. The IscS/CyaY complex affinity increases considerably from $K_d = 23 \pm 3$ µM to $K_d = 35 \pm 6$ nM when IscU is present [35]. In saturating conditions of IscU over IscS, the fractional occupancy of CyaY on IscS/IscU varies from 0 for [CyaY] = 0 µM to 100% for [CyaY] = 10 µM (Table 4.2). However, for the non-saturating conditions of the present study, an occupancy of IscU on IscS of ~53% is predicted for [IscS] = 1 µM and [IscU] = 2 µM. The chosen IscS and IscU concentrations are in order to replicate the occupancy of ~50% determined for IscU on IscS in the cellular environment of *E. coli* (unpublished data obtained from a collaboration work). For comparison purposes, Table 4.2 shows the predicted fractional occupancy values of CyaY on IscS in the absence of IscU. Even at the maximum concentration used for CyaY, occupancy values as low as ~22% are predicted. The coexistence of IscU bound and unbound to IscS is thus an intermediate scenario between the ones described in terms of fractional occupancy. In addition, the dissociation constant described for the formation of IscS/IscU refers to CyaY-free conditions when, in fact, the formation of IscS/IscU complex is also enhanced by the presence of CyaY [35]. This three-way interaction between IscS, IscU and CyaY increases the complexity of fractional occupancy estimation for any of the proteins involved.

The mixed character of the present inhibition mechanism may possibly result from the simultaneous occurrence of free and IscU-bound IscS. Since CyaY binds differently to each of these species, the predominantly specific or catalytic character of the inhibition mechanism might be primarily determined by the concentration of CyaY. Overall, these results support the theory of CyaY as a modifier and gatekeeper of IscS activity [14], and that CyaY exerts its inhibitory effect after the substrate is bound to IscS.

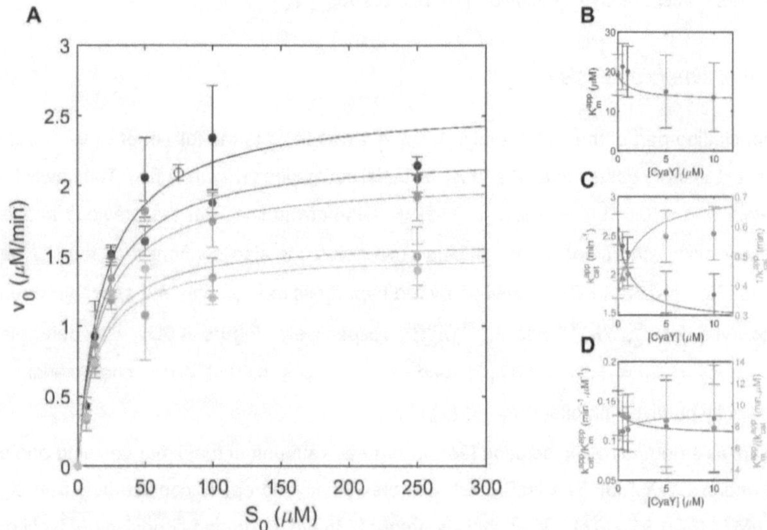

Figure 4.6. Final GMM analysis and parameter retrieval. (A) Symbols and error bars: means and standard deviations of measured v_0 values (color-coded as in Figure 4.3). Solid lines: nonlinear fitting of the GMM equation (Eq. 4.1) to the full v_0 vs S_0 dataset (selected experimental data, closed markers). Fitted result: $\alpha = 0.671 \pm 0.183$, $\beta = 0.538 \pm 0.054$, $K_I = 1.84 \pm 0.89 \, \mu M$. (B-D) Dependence of apparent kinetic parameters (B) K_m^{app}, (C) k_{cat}^{app} (and reciprocal $1/k_{cat}^{app}$) and (D) k_{cat}^{app}/K_m^{app} (and reciprocal K_m^{app}/k_{cat}^{app}) on the modifier concentration. Symbols and error bars: apparent parameters and standard errors for 95% confidence level; solid lines: representation of Eqs. 4.2a-4.2c and Eqs. 4.3a-4.3b using fitted parameters.

Table 4.2. Fractional occupancy of CyaY on IscS (1 µM) in the absence of IscU (K_d = 23 ± 3 µM [35]) or under saturating conditions of IscU over IscS (K_d = 35 ± 6 nM [35]). Fractional occupancy was determined considering a reversible 1:1 complex between a protein P and ligand L ($P + L \rightleftharpoons PL$) [36].

[CyaY] (µM)		0	0.5	1	5	10
Fractional occupancy of CyaY on IscS	Saturating IscU	0%	47%	83%	99%	100%
	Absence of IscU	0%	1%	3%	12%	22%

4.5. <u>Conclusions</u>

In the present work, the influence of CyaY as a modulator of the desulfurase activity of IscS was studied in the presence of IscU. The kinetic mechanism behind this modulation was assessed by application of the GMM. For the cell-like conditions that were studied, CyaY was identified as a hyperbolic mixed inhibitor with predominantly catalytic character, typical of an allosteric modulator. The change in free energy due to allosteric coupling is −0.99 kJ/mol [18, p.68]. The inhibitory activity of CyaY does not require the presence of iron, even if increasing concentrations of iron potentiate the CyaY effect [14]. As an inhibitor of Fe-S cluster formation, CyaY is confirmed to target the desulfurase step; further experiments are needed in order to study the influence of CyaY in the step of Fe-S cluster assembly. The described methodology may also be applied in the future to determine the kinetic mechanisms by which other IscS interactors such as ferredoxin (Fdx) [13] or YfhJ [37,38] act over Fe-S cluster biogenesis. Elucidating these mechanisms at the molecular scale should contribute for a better understanding of the pathophysiology of FRDA in view of possible pharmacological treatments of this neurodegenerative disease.

<u>References</u>

1.   Campuzano, V., *et al.* (1996), *Friedreich's ataxia: autosomal recessive disease caused by an intronic GAA triplet repeat expansion.* Science. 271(5254):1423-7.

2.   Campuzano, V., *et al.* (1997), *Frataxin is reduced in Friedreich ataxia patients and is associated with mitochondrial membranes.* Human Molecular Genetics. 6(11):1771-80.

3.   Pandolfo, M. and Pastore, A. (2009), *The pathogenesis of Friedreich ataxia and the structure and function of frataxin.* Journal of Neurology. 256 Suppl 1:9-17.

4.   Muhlenhoff, U., *et al.* (2002), *The yeast frataxin homolog Yfh1p plays a specific role in the maturation of cellular Fe/S proteins.* Human Molecular Genetics. 11(17):2025-36.

5.   He, Y., *et al.* (2004), *Yeast frataxin solution structure, iron binding, and ferrochelatase interaction.* Biochemistry. 43(51):16254-62.

6.   Foury, F. and Cazzalini, O. (1997), *Deletion of the yeast homologue of the human gene associated with Friedreich's ataxia elicits iron accumulation in mitochondria.* FEBS Letters. 411(2-3):373-7.

7.   Lesuisse, E., *et al.* (2003), *Iron use for haeme synthesis is under control of the yeast frataxin homologue (Yfh1).* Human Molecular Genetics. 12(8):879-89.

8.   Yoon, T. and Cowan, J.A. (2004), *Frataxin-mediated iron delivery to ferrochelatase in the final step of heme biosynthesis.* Journal of Biological Chemistry. 279(25):25943-6.

9.   Gerber, J., Muhlenhoff, U., and Lill, R. (2003), *An interaction between frataxin and Isu1/Nfs1 that is crucial for Fe/S cluster synthesis on Isu1.* EMBO Reports. 4(9):906-11.

10.   Ramazzotti, A., Vanmansart, V., and Foury, F. (2004), *Mitochondrial functional interactions between frataxin and Isu1p, the iron-sulfur cluster scaffold protein, in Saccharomyces cerevisiae.* FEBS Letters. 557(1-3):215-20.

11.   Layer, G., *et al.* (2006), *Iron-sulfur cluster biosynthesis: characterization of Escherichia coli CYaY as an iron donor for the assembly of [2Fe-2S] clusters in the scaffold IscU.* Journal of Biological Chemistry. 281(24):16256-63.

12.   Yoon, T. and Cowan, J.A. (2003), *Iron-sulfur cluster biosynthesis. Characterization of frataxin as an iron donor for assembly of [2Fe-2S] clusters in ISU-type proteins.* Journal of the American Chemical Society. 125(20):6078-84.

13.   Yan, R., *et al.* (2013), *Ferredoxin competes with bacterial frataxin in binding to the desulfurase IscS.* Journal of Biological Chemistry. 288(34):24777-87.

14.   Adinolfi, S., *et al.* (2009), *Bacterial frataxin CyaY is the gatekeeper of iron-sulfur cluster formation catalyzed by IscS.* Nature Structural & Molecular Biology. 16(4):390-6.

15.   Kim, J.H., *et al.* (2015), *Tangled web of interactions among proteins involved in iron-sulfur cluster assembly as unraveled by NMR, SAXS, chemical crosslinking, and functional studies.* Biochimica et Biophysica Acta. 1853(6):1416-28.

16.   Botts, J. and Morales, M. (1953), *Analytical description of the effects of modifiers and of enzyme multivalency upon the steady state catalyzed reaction rate.* Transactions of the Faraday Society. 49(0):696-707.

17.   Roche, B., *et al.* (2013), *Iron/sulfur proteins biogenesis in prokaryotes: formation, regulation and diversity.* Biochimica et Biophysica Acta. 1827(3):455-69.

18.   Baici, A., (2015), *Kinetics of Enzyme-Modifier Interactions*, 1st ed. Springer.

19.   Morales, M.F. (1955), *If an Enzyme-Substrate Modifier System Exhibits Non-competitive Interaction, then, in General, its Michaelis Constant is an Equilibrium Constant.* Journal of the American Chemical Society. 77(15):4169-4170.

20. Cárdenas, M.L. and Cornish-Bowden, A. (1989), *Characteristics necessary for an interconvertible enzyme cascade to generate a highly sensitive response to an effector.* Biochemical Journal. 257(2):339.

21. Nair, M., *et al.* (2004), *Solution structure of the bacterial frataxin ortholog, CyaY: mapping the iron binding sites.* Structure. 12(11):2037-48.

22. Adinolfi, S., *et al.* (2002), *A structural approach to understanding the iron-binding properties of phylogenetically different frataxins.* Human Molecular Genetics. 11(16):1865-77.

23. Ramelot, T.A., *et al.* (2004), *Solution NMR Structure of the Iron–Sulfur Cluster Assembly Protein U (IscU) with Zinc Bound at the Active Site.* Journal of Molecular Biology. 344(2):567-583.

24. Kim, J.H., *et al.* (2009), *Structure and dynamics of the iron-sulfur cluster assembly scaffold protein IscU and its interaction with the cochaperone HscB.* Biochemistry. 48(26):6062-6071.

25. Prischi, F., *et al.* (2010), *Of the vulnerability of orphan complex proteins: The case study of the E. coli IscU and IscS proteins.* Protein Expression and Purification. 73(2):161-166.

26. Siegel, L.M. (1965), *A direct microdetermination for sulphide.* Anal Biochem. 11:126-32.

27. Bridwell-Rabb, J., *et al.* (2012), *Effector Role Reversal during Evolution: The Case of Frataxin in Fe–S Cluster Biosynthesis.* Biochemistry. 51(12):2506-2514.

28. Cook, A.G., Tolliver, R.M., and Williams, J.E. (1994), *The Blue Bottle Experiment Revisited: How Blue? How Sweet?* Journal of Chemical Education. 71(2):160.

29. Zhu, A., Romero, R., and Petty, H.R. (2010), *A sensitive fluorimetric assay for pyruvate.* Analytical Biochemistry. 396(1):146-51.

30. Shi, R., *et al.* (2010), *Structural basis for Fe-S cluster assembly and tRNA thiolation mediated by IscS protein-protein interactions.* PLOS Biology. 8(4):e1000354.

31. di Maio, D., *et al.* (2017), *Understanding the role of dynamics in the iron sulfur cluster molecular machine.* Biochimica et Biophysica Acta (BBA) - General Subjects. 1861(1, Part A):3154-3163.

32. Stevens, R., Stevens, L., and Price, N.C. (1983), *The stabilities of various thiol compounds used in protein purifications.* Biochemical Education. 11(2):70-70.

33. Baici, A. (1981), *The Specific Velocity Plot.* European Journal of Biochemistry. 119(1):9-14.

34. Cupp-Vickery, J.R., Urbina, H., and Vickery, L.E. (2003), *Crystal Structure of IscS, a Cysteine Desulfurase from Escherichia coli.* Journal of Molecular Biology. 330(5):1049-1059.

35. Prischi, F., *et al.* (2010), *Structural bases for the interaction of frataxin with the central components of iron–sulphur cluster assembly.* Nature Communications. 1(1):95.

36. Podjarny, A., Dejaegere, A.P., and Kieffer, B., (2011), *Biophysical Approaches Determining Ligand Binding to Biomolecular Targets: Detection, Measurement and Modelling*, 1st ed. RSC Biomolecular Sciences. Royal Society of Chemistry, pp.118.

37. Adinolfi, S., *et al.* (2018), *The molecular bases of the dual regulation of bacterial iron sulfur cluster biogenesis by CyaY and IscX*. Frontiers in Molecular Biosciences. 4:97.

38. Pastore, C., *et al.* (2006), *YfhJ, a Molecular Adaptor in Iron-Sulfur Cluster Formation or a Frataxin-like Protein?* Structure. 14(5):857-867.

Chapter 5.

Major improvements in robustness and efficiency during the screening of novel enzyme effectors by the 3-point kinetics assay

The content of the present chapter constitutes an original research article in preparation (Article IV):

Pinto, M.F., Silva, A., Figueiredo, F., Pombinho, A., Pereira, P.J.B., Macedo-Ribeiro, S., Rocha, F., Martins, P.M., Major improvements in robustness and efficiency during the screening of novel enzyme effectors by the 3-point kinetics assay.

5.2. <u>Introduction</u>

The study of enzyme kinetics commonly focuses the initial reaction phases during which constant rate conditions are verified irrespective of the degree of substrate conversion. During the so-called steady-state period, the measured initial reaction rate (v_0) is related with the initial substrate concentration (S_0) according to the Michaelis-Menten (MM) equation [1]. The early stages of reactions are of additional importance for the screening of novel enzyme inhibitors since the percentage inhibition computed in terms of apparent reaction rates is known to decrease over time [2,3]. Moreover, end-point assays performed within the period of constant v_0 are associated to minimal costs per screened compound when large numbers (> 10,000) of chemical compounds are being tested [2,4].

With the advent of robotics and sample miniaturization systems, continuous monitoring of multiple reactions became possible through the use of highly automated workstations [5]. While covering different phases of the reaction, full progress curve analysis has the potential to increase the screening efficiency through the detection of enzyme modulation effects that are either slow to manifest or masked at high substrate concentrations [4,6,7]. Deconvoluting this useful information from the occurrence of, e.g., unaccounted enzyme inactivation, instrumental drift and compound interference is, however, difficult even considering separated timescales [8,9] (see Chapters 2 and 3). In a previous attempt to simplify high-throughput screenings (HTS) of dynamic modulation effects, the theoretical relationship between the half maximal inhibitory concentration (IC_{50}) with the substrate conversion was established assuming first-order kinetic models [10]. In another example, end-point enzymatic assays were optimized in order to measure the reaction conversion at the point of maximum difference in product concentration between control and competitively inhibited reactions [7].

A minimum of 3 point readouts are required to probe the initial, intermediate and final phases of screening reactions [11] (Figure 5.1A). More frequent readings are technologically possible [12,13], although their implementation in the phase of primary screening is avoided in order to keep the assay cost to a minimum [4]. When conditions of large substrate concentration cannot be adopted, the throughput level of end-point assays is greatly limited by the duration of the constant rate period, at the end of which different estimates of v_0 start to be produced [14,15] (Figure 5.1B).

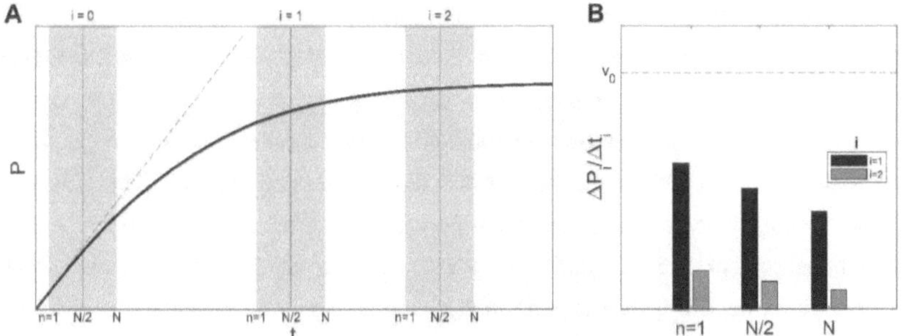

Figure 5.1. Numerical example showing possible configurations of endpoint and 3-point HTS assays. (A) The simulated enzymatic reaction (solid line) is assumed to run simultaneously in a total of $n = N$ wells. The product concentration P in each well can be probed 3 times ($i = 0, 1, 2$) over periods of time δt (shaded areas) whose duration is determined by the technological throughput level as well as by the value of N. The slope of the dashed line indicates the true initial rate v_0. (B) Changeable estimates of apparent rates $(P_i - P_0)/(t_i - t_0)$ will be produced unless the δt interval is considerably shorter than the full timespan.

Compound-induced changes of background signal can be automatically normalized by adopting a two-point strategy to measure the increase of product concentration ($\Delta P_i = P_i - P_0$). This advantage, which is shared by kinetic-mode assays, obviates the design of counter-assays for compound interference identification [16-18]. Nevertheless, increasing the reaction monitoring to 2 or more instants does not solve the problem of time-dependent readouts unless a renewed theoretical framework is adopted.

After proposing the closed-form mathematical solution of single substrate, single active-site enzymatic mechanisms [19,20] (see Introduction, Section II, and Chapter 1, Section 1.5.1), our group developed the linearization method (LM) for the detection of hidden assay interferences [9] (see Chapter 2, Section 2.4). Next, we show that the LM principles can be applied to the HTS detection of «interferences» caused by candidate enzyme effectors. The assets of the new method are demonstrated by combining numerical simulations with a practical example of compound screening for modulators of the deubiquinating activity of polyglutamine-expanded ataxin-3 (Atx-3 77Q), a protein associated to spinocerebellar ataxia type 3, also known as Machado-Joseph disease [21].

5.3. <u>Materials and Methods</u>

5.3.1. Protein Production

Atx-3 77Q was expressed and purified as previously described [22-24]. Briefly, pDEST17-ATX3(77Q) plasmid was transformed into *E. coli* BL21(DE3)-SI cells (Life Technologies, Carlsbad, CA, USA). For protein expression, cells were first grown at 37 °C, 180 rpm, in LB medium without NaCl in the presence of 100 µg/mL of ampicillin and 0.4% (w/v) glucose, until optical density (OD) at 600 nm reached 0.6-0.8, then cooled down to 30 °C, and induced with 300 mM NaCl for 3 hours. Cell pellets were harvested and resuspended in buffer A (20 mM sodium phosphate pH 7.5, 500 mM NaCl, 2.5% (v/v) glycerol, 20 mM imidazole) containing 100 mg/L lysozyme. Cell lysis was performed by stirring 1 hour on ice in the presence of 0.02 mg/mL DNase, 1 mM $MgCl_2$ and 1 mM PMSF. After centrifugation, the supernatant was loaded onto a Ni^{2+}-charged HisTrap column (GE Healthcare Life Sciences) equilibrated in buffer A, and protein was eluted with a gradient of imidazole (50 mM, 250 mM and 500 mM). Atx-3 77Q was further purified using a HiPrep 26/60 Sephacryl S-200 HR column (GE Healthcare Life Sciences) equilibrated in buffer (20 mM sodium phosphate pH 7.5, 200 mM NaCl, 5% (v/v) glycerol, 2 mM EDTA, 1 mM DTT). Protein purity was checked using SDS-PAGE after each purification step and for the final product. Final protein stock concentration was determined by measuring the absorbance at 280 nm using the molar absorption coefficient of 31650 $M^{-1}cm^{-1}$.

5.3.2. Enzymatic Assay

A modified version of the method described by Burnett *et al.* (2003) [25] was employed to monitor the catalysis of the fluorogenic substrate Ubiquitin-AMC (Ub-AMC, Boston Biochem) to fluorescent 7-amino-4-methylcoumarin (AMC) by Atx-3 77Q. Enzymatic reactions were performed in 50 mM HEPES, 0.5 mM EDTA, pH 7.5 (prepared at 20 °C), with 0.1 mg/ml BSA, 5% (v/v) glycerol and 10 mM DTT at 37 °C using 384-well microplates (Corning®, Low Flange, Black, Flat Bottom, Polystyrene) and a total reaction volume of 50 or 30 µL per well. A range of Ub-AMC concentrations lower than 1 µM was used, as per recommendation of the substrate manufacturer. Total enzyme concentration ranged from 0.25 to 0.5 µM. Reactions were started by addition of protein, and fluorescence was monitored at 460 nm (390 nm excitation) using a HIDEX CHAMELEON V plate reader (Turku, Finland). To avoid evaporation, the reaction mixture in each well was overlaid with liquid paraffin (30 µL).

5.3.3. Screening procedures

Potential modulators of Atx-3 77Q activity were screened from 1280 FDA-approved chemical compounds contained in the Prestwick Chemical Library® (Prestwick Chemical). An automated bulk dispenser (Multidrop Combi Thermo Scientific) was used to fill 384-well microplates with 15 µL of 0.75 µM Ub-AMC. An automated liquid handler (JANUS Automated Workstation, PerkinElmer) equipped with pin tool replicators (V&P Scientific) coupled to an Modular Dispense Technology (MDT) head was used to add 0.1 µL of test compounds (from a 1 mM stock) or DMSO (for controls), after which reactions were started by adding 15 µL of 0.70 µM Atx-3 77Q to each well (reaction volume: 30 µL). The MDT head of the automatic liquid handler equipped with 96 tips was used to dispense 30 µL liquid paraffin to the final reaction mixtures. A total of four 384-well microplates were filled, each plate testing 320 chemical compounds (1 compound per well) and running 32 control reactions in the presence of DMSO (DMSO control reactions). The final reaction mixture contained 0.35 µM Atx-3 77Q, 0.375 µM Ub-AMC and 3.33 µM test compound/DMSO. Assay robustness was evaluated separately for each microplate in terms of the median Z-factor [26,27]:

$$Z = 1 - \frac{3\sigma_s - 3\sigma_c}{|\tilde{x}_s - \tilde{x}_c|} \tag{5.1}$$

where $\tilde{x}$ and σ represent median and standard deviation, index s refers to the analyzed sample, corresponding to 320 test reactions + 32 DMSO control reactions, and index c refers to 32 blank reactions corresponding to negative controls. Two different normalization methods were investigated before and after the application of three exclusion criteria described below.

Specifically, values of Z-factor were computed based on (i) initial reaction rates determined over a period of ~1h duration, and on (ii) apparent reaction rates measured by the 3-point method.

5.3.4. Criteria for compound selection

According to the LM, the following linear relationship between the modified reaction coordinates $\Delta P_i/\Delta t_i$ and $-\ln(1 - \Delta P_i/\Delta P_\infty)/\Delta t_i$ is established under typical conditions of large substrate excess over enzyme, and independently of which reaction timeframes or initial substrate concentrations are considered [9] (see Chapter 2, Section 2.4):

$$\frac{\Delta P_i}{\Delta t_i} = V^{app} - K_m^{app}\left[-\frac{\ln\left(1 - \frac{\Delta P_i}{\Delta P_\infty}\right)}{\Delta t_i}\right] \tag{5.2}$$

where ΔP_∞ corresponds to the product concentration increase measured at the end of the reaction, and K_m^{app} and V^{app} are apparent kinetic constants. The values of K_m^{app} and V^{app} expected for interference-free control reactions correspond to the Michaelis constant (K_m) and limiting rate (V), respectively. Added compounds inhibiting or promoting enzyme activity will change the location of the LM coordinates to a region below (for inhibitors) or above (for activators) the straight line of slope $-K_m$ and vertical axis intercept V that is defined by Eq. 5.2 (Appendix Section 5.6.1, Figure A5.1).

The proposed method of enzyme modifier detection relies on the measurement of product concentration P_i (or, alternatively, substrate concentration S_i) in three distinct moments (indexes $i = 0, 1, 2$). When the effects of the screening compounds on the calibration curve are not known beforehand, indirect monitoring of product concentration is performed using signal readout values F_i instead of P_i. Possible changes in background signal are normalized by using ΔF_i differences in the estimation of midpoint-centered ($i = 1$) LM coordinates:

$$X_1 = -\frac{\ln\left(1 - \frac{\Delta F_1}{\Delta F_2}\right)}{\Delta t_1} \tag{5.3a}$$

$$Y_1 = \frac{\Delta F_1}{\Delta t_1} \tag{5.3b}$$

Preferably, the third-point location should coincide with the final plateau reached at the end of the reaction thereby admitting the equivalence between $\Delta F_1/\Delta F_2$ and $\Delta P_1/\Delta P_\infty$ ratios. As the third-point becomes located further behind the reaction endpoint, the resolution of the method is expected to decrease (Appendix Section 5.6.1, Figure A5.3). The time- and S_0-independent LM criterion for compound selection is based on the difference between obtained and predicted X_1 or Y_1 values. The stronger the effect of the modulator, the larger the differences ΔX_1 or ΔY_1 will be. During the screening of Atx-3 77Q effectors, the resolution of the method is computed in terms of the Y_1 difference:

$$\Delta Y_1 = Y_1 - (V - K_m X_1)\kappa \tag{5.4}$$

where κ is the known proportionality factor converting product concentration units into F_i units in the absence of test compounds. If all test reactions can be probed simultaneously (i.e., if $\delta t \ll \Delta t_i$ in the example of Figure 5.1), previous knowledge of the control parameters (K_m and V) is not required. In such cases, distribution histograms are sufficient to statistically select higher (for activators) and lower (for inhibitors) values of X_1 or Y_1.

During compound screening for modulators of Atx-3 77Q activity, three exclusion criteria are adopted to eliminate test reactions affected by random experimental error and artifacts. The first criterion accounts for evident deviations from the product build-up trend caused, for example, by sudden fluorescence quenching. To pass this criterion, the following condition has to be verified:

$$0 < \Delta F_1 \leq \Delta F_2 \tag{5.5}$$

The enzymology principles used to define the other two criteria are described in detail in Appendix Section 5.6.2. The second validation criterion is derived taking as reference the limit case of $K_m^{app} \approx 0$ for which product concentration increases linearly with time:

$$\frac{\Delta t_1}{\Delta t_2} \leq \frac{\Delta F_1}{\Delta F_2} \tag{5.6}$$

In the other limit case of $K_m^{app} \gg S_0$, product concentration is expected to follow an asymptotic exponential growth [20] (see Chapter 1, Section 1.4.1). With the values of $\Delta F_1/\Delta F_2$ and Δt_1, a time constant for exponential P_i-increase is defined setting the physical limits for the maximum $\Delta t_1/\Delta t_2$ ratio. While more stringent than Eq. 5.6 for high $\Delta F_1/\Delta F_2$ ratios, these limits are used to establish the third validation criterion:

$$\frac{\Delta t_1}{\Delta t_2} \leq -\ln\left(1 - \frac{\Delta F_1}{\Delta F_2}\right)\left(\frac{1}{\Delta F_1/\Delta F_2} - 1\right) \tag{5.7}$$

5.4. <u>Results</u>

The improvements in robustness brought forward by the 3-point LM assay are illustrated with the screening of 1280 FDA-approved chemical compounds for possible modulators of the deubiquinating activity of Atx-3 77Q. This is a proposedly small chemical library that allows for kinetic-mode monitoring of all tested reactions running in four 384-well microplates (Figure 5.2). The occurrence of spurious phenomena eventually suggested by the 3-point positions can thus be evaluated against what is reported by the full progress curves. In all plates, control-reaction conversions of ~0%, ~60% and ~99.9% were chosen for the 3-point positions along the time-courses. The number of excluded compounds greatly changes from plate to plate as a possible consequence of the different chemical nature of tested molecules, but also owing to inherent random and systematic errors arising, for example, from differences in the effective concentration of active sites of Atx-3 77Q, which is an aggregation-prone protein [23].

The majority of excluded curves is identified upon the application of the first filter accounting for evident drops in AMC fluorescence caused by quenching interference, inner filter effects, light scattering, etc. (solid red lines in Figure 5.2). Examples of supra-linear trends failing to comply with Eq. 5.6 are not observed, whereas the third exclusion criterion (Eq. 5.7) does eliminate further unreliable readouts (dashed red lines in Figure 5.2). Unsurprisingly, the number of excluded results increases as the second point is positioned nearer to the third point (Table 5.1). Higher $\Delta F_1/\Delta F_2$ ratios increase the risk of false negatives as the result of possible deviations from physically acceptable trends originated by, e.g., signal noise (Appendix Section 5.6.2); by adopting $\Delta F_1/\Delta F_2$ ratios below 0.5, the safety margin for exclusion increases, yet, the false positive rates in subsequent phases of hit detection are also expected to increase. All of the suspicious curves that were eliminated using a midpoint threshold of ~60% control-reaction conversion (Figure 5.2), are indeed affected by systematic drops in fluorescence signal during, at least, 20% of the time over which the reactions are monitored (Table 5.1, last column).

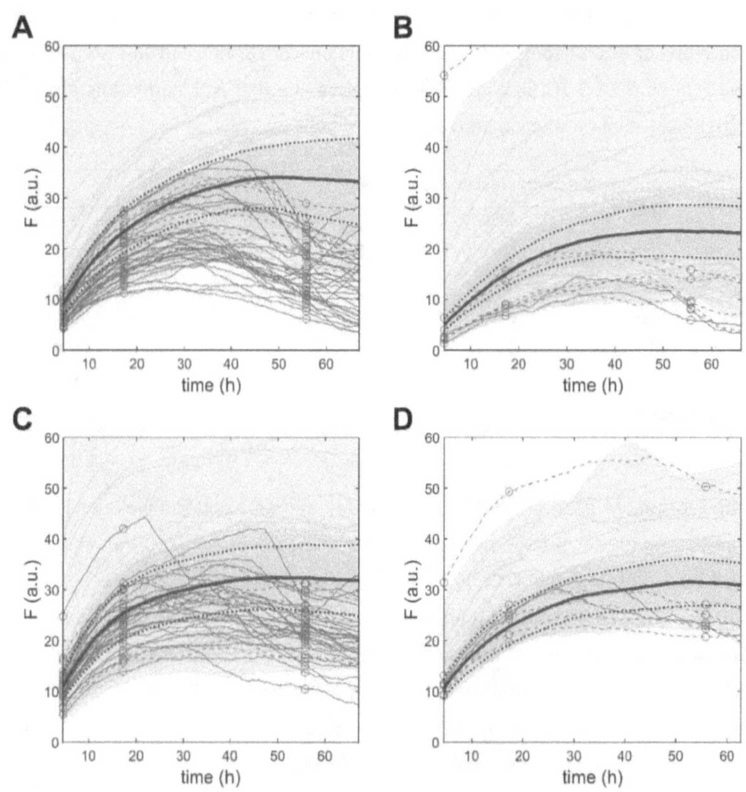

Figure 5.2. Validation criteria applied to the screening of drug repurposing compounds for possible modulators of the enzymatic activity of Atx-3 77Q. (A-D) Zoomed progress curves showing mean and standard deviation values of control reaction readouts (solid and dashed blue lines, respectively); excluded (red lines) and validated (gray lines) screening reactions; and the 3-point positions (symbols) along the excluded progress curves. Full scale graphs are given as Appendix Section 5.6.4, Figure A5.6. Filters to eliminate spurious results are successively applied based on Eqs. 5.5 (solid red lines), 5.6 (no results excluded) and 5.7 (dashed red lines). Different panels present the results obtained in microplates (A) #1, (B) #2, (C) #3, (D) #4. Areas shaded in light gray represent intervals where validated test reactions can be found.

Table 5.1. Impact of the midpoint position on the total number of excluded results. The reference case of ~60% control-reaction conversion corresponds to the sum of excluded results in Figures 5.2A-5.2D. In all considered cases, the first and third points are located at ~0% and ~99.9% control-reaction conversions.

Control reaction conversion (%)	Criterion 1 (Eq. 5.5)	Criterion 2 (Eq. 5.6)	Criterion 3 (Eq. 5.7)	Total excluded	Percentage of success[b]
	Number of excluded results (out of 1404)				
60[a]	69	0	29	98	100
70	123	0	175	298	92.6
50	36	0	9	45	100
40	18	0	2	20	100

[a]Chosen midpoint threshold
[b]Results are confirmed as bad outliers when the fluorescence signal persistently drops over periods longer than 20% of the full time-course

The AMC fluorophore used in the Atx-3 77Q assay has an excitation wavelength in the UV range, which also excites a large number of screening compounds [28]. Values of median Z-factor well below the practical limit of 0.5 confirm that the accuracy of this drug repurposing screening would not be acceptable unless robust exclusion criteria were used to identify assay artifacts (Table 5.2). The major improvements in robustness reported in Table 5.2 are achieved through the elimination of bad outliers (compare the last two columns of the table) but also by adopting better normalization procedures than those conventionally adopted for endpoint assays (compare the v_0- and the $Y_1/\Delta F_2$-based Z-factors).

Table 5.2. Median Z-factors (Eq. 5.1) calculated for each microplate before and after the LM-based exclusion criteria are applied. Initial rate and ΔF_2-normalized Y_1 values are used as assay readouts.

Plate #	v_0 Readouts		$Y_1/\Delta F_2$ Readouts[a]	
	Before	After	Before	After
1	-4.52	-4.08	-13.23	0.41
2	-0.49	-0.45	0.28	0.45
3	-1.59	-0.25	-0.18	0.50
4	-0.01	0.17	0.33	0.49

[a]The ΔF_2 values used for negative controls correspond to mean ΔF_2 values obtained during DMSO control reactions

Initial rate analysis of control reactions revealed inter-plate differences in enzymatic activity despite the fact that the same molar concentration of Atx-3 77Q is used across microplates. Such variability is likely due to variations in the oligomerization state of this polyglutamine-expanded protein, whose characterization in terms of the kinetic parameters K_m and V was performed adopting lower enzyme concentration than during compound screening (Appendix Section 5.6.3.2, and Section 5.6.4, Figure A5.7). The resulting LM straight lines (solid lines in Figures 5.3A-5.3D) underestimate most of the X_1 and Y_1 values obtained for each microplate (symbols in Figures 5.3A-5.3D), even after the correction of parameter V by the effective enzyme concentration. While systematically affecting the differences between obtained and predicted Y_1 values (Eq. 5.4), this variability has no major influence on the 3-point method of compound selection. Illustrating this, inter-plate variations in Y_1 estimates (Figure 5.4A) tend to vanish when the results are represented as ΔY_1 distributions (Figure 5.4B). Since the 3 points are probed at practically simultaneous time instants, within and across microplates, the dispersion of Y_1 (or X_1) values is not justified by significant variations in the reaction periods considered in each well. Should these variations have occurred, their normalization across microplates would also be possible by representing frequency distributions of ΔY_1 values (see for details the numerical examples in Appendix Section 5.6.1, Figure A5.1 and Figure A5.2).

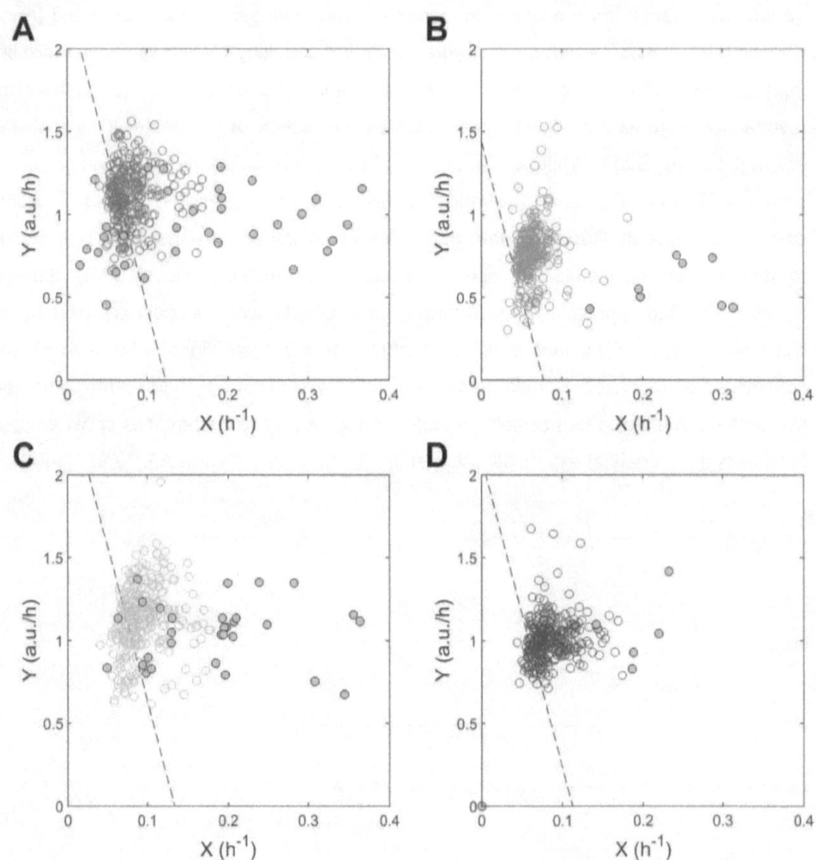

Figure 5.3. Zoomed LM representation of screening reactions. Full scale graphs are given as Appendix Section 5.6.4, Figure A5.8. (A-D) Symbols: intermediate LM coordinates computed for validated (open symbols) and excluded (closed symbols in gray) results using Eq. 5.3; for graphical representation purposes, argument of the logarithm in Eq. 5.3a is determined as its module. Dashed Lines: theoretical LM straight lines computed from Eq. 5.2 using the values of K_m and V obtained during Atx-3 77Q assay characterization (Appendix Section 5.6.3.2, and Section 5.6.4, Figure A5.7); parameter V is corrected by a factor of (A) 0.94, (B) 0.59, (C) 1.02, (D) 0.86 accounting for the different Atx-3 77Q activities measured at 0.375 μM Ub-AMC during assay characterization and during the control reactions in each microplate (Appendix Section 5.6.3).

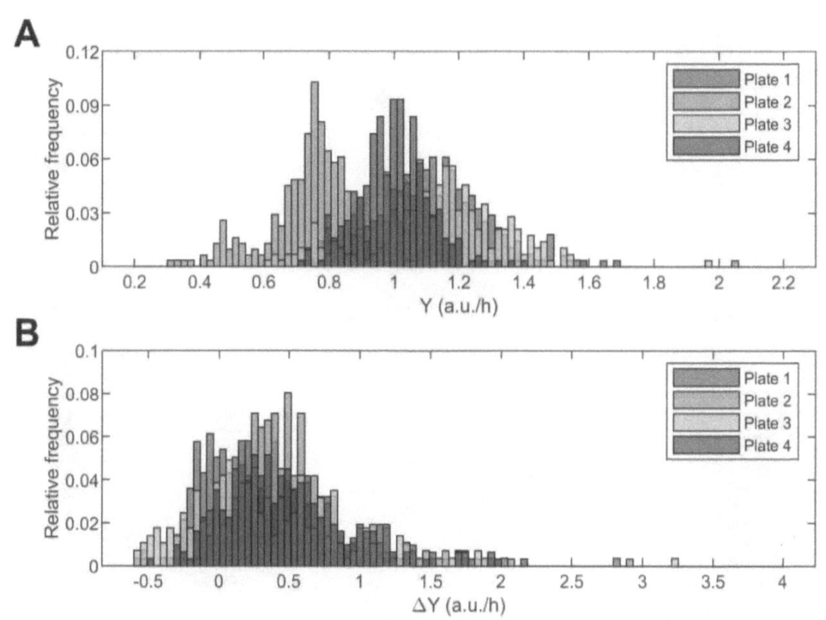

Figure 5.4. Histograms showing the frequency distribution of midpoint LM coordinates obtained for each microplate. (A-B) Only validated data are used to compute (A) Y_1 values from Eq. 5.3b, and (B) ΔY_1 differences from Eq. 5.4. The usage of ΔY_1 as compound selection criterion normalizes the variability associated to the effective concentration of active enzyme.

5.5. Discussion

The 3-point method of kinetic analysis answers the current demand for new breakthroughs in the discovery of inhibitors and activators for targets of interest in pharmaceutical research [29-31]. Its first principles are those of the linearization method recently proposed by us as a new tool to detect enzymatic assay interferences [9] (see Chapter 2), with the main difference that LM curves are now used to detect «interferences» provoked by candidate enzyme effectors. Using the screening for Atx-3 77Q modulators as a practical example, major improvements in robustness are achieved through judicious elimination of bad outliers, by adopting advanced normalization methods, and increasing the assay resilience to systematic and random variability. Simple numerical examples are given in Figure 5.5 to illustrate another way by which the 3-point LM assay can improve HTS efficiency, namely by increasing the detection of true positive hits. Mechanisms of specific (or competitive), catalytic (or uncompetitive) and slow-binding inhibition are simulated assuming $S_0 = 10 K_M$.

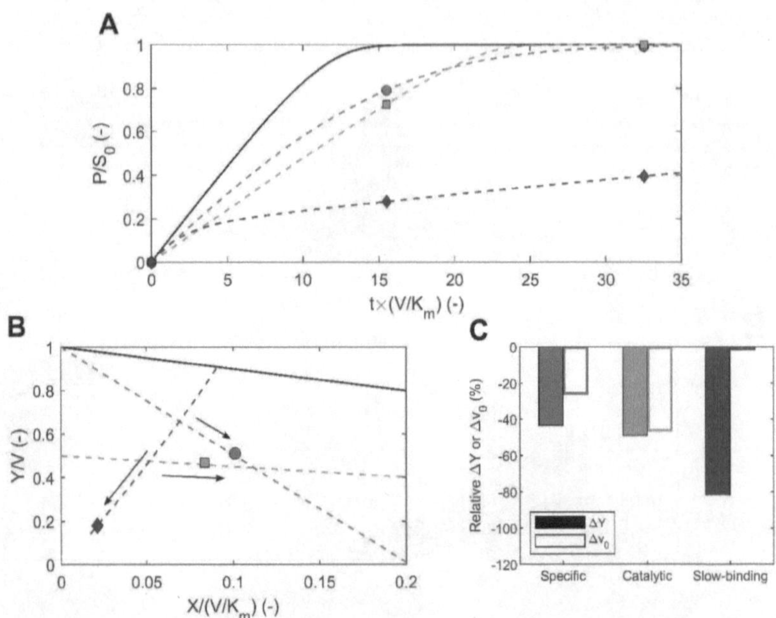

Figure 5.5. Numerical simulations of different enzyme modifier mechanisms assuming $S_0 = 10K_M$. Further details concerning numerical methods are given in Appendix Sections 5.6.3 and 5.6.5 (Table A5.1). (A) Lines: S_0-normalized time-courses for control (solid line) and test reactions of (dashed green line) catalytic, (dashed red line) specific and (dashed blue line) slow-binding inhibition. Symbols: possible selection of reaction coordinates by the 3-point method. (B) Lines: LM representations of the time-courses in (A). Quasi-equilibrium conditions are assumed for specific and catalytic inhibition; Eq. 5.3 extended to the full time-course is used to compute the LM curve in the case of slow-binding inhibition. Symbols: linearized midpoint coordinates corresponding to the 3-point selection. Arrows indicate progress in time. (C) Full bars: resolution of the 3-point method given as ΔY_1 differences (Eq. 5.4) normalized by the expected value of Y_1 for control reactions. Open bars: relative differences between values v_0 determined from the onset of test and control progress curves.

Substrate concentrations as high as $10K_M$ are beneficial to increase the signal-to-background ratio and to detect uncompetitive inhibitors, yet, values of $S_0 \approx K_M$ are generally preferred in HTS practice as a compromise to warrant good responsiveness to competitive inhibitors as well [3,30,32]. In addition to a marked increase in sensitivity to competitive inhibitors at high S_0 values, the numerical simulations in Figure 5.5 demonstrate that the 3-point screening method still marginally increases the detection limits for uncompetitive inhibitors. Therefore, in the light of the new methodology, high substrate concentrations can henceforth be adopted without further constraints than the practical limits imposed by substrate solubility and the desired overall duration of the assay. The LM straight lines describing the catalytic and specific mechanisms in Figure 5.5B are simulated assuming quasi-equilibrium conditions that do not apply to the case of slow-binding inhibition [33]. Transient enzyme modifier mechanisms are particularly difficult to detect solely based on initial rate measurements. Conversely, the 3-point method is capable of greatly improving the assay sensitivity to slow-binding inhibitors provided that sufficiently long Δt_1 periods are adopted (Figure 5.5C).

The cases depicted in Figure 5.5 are necessarily limited in number, as many other mechanisms of inhibition, inactivation, activation, etc., could be studied for various combinations of substrate, enzyme, and enzyme modifier concentrations. Independently of which scenario is considered, the LM remains highly sensitive to minor kinetic variations [9] (see Chapter 2), whereas the precursory application of robust validation criteria prevents false hit proliferation. Whenever possible, initial substrate concentrations in the order of magnitude of $S_0 \gg K_m$ should be adopted so as to warrant high efficiencies of the 3-point method during exclusion and selection of HTS results. Although no fixed procedure is imposed for the 3-point selection, narrow ranges of signal readout variation are not recommended owing to the noise amplification originated by the logarithm term in the definition of X_1 (Eq. 5.3a). Consequently, as a rule of thumb, the first and third points may coincide with beginning and conclusion of the assay, while the second point may correspond to instantaneous substrate concentrations of $\sim K_M$ (taking the control reaction progress curves as reference). The suggested midpoint location may change to earlier or later moments depending on whether enzyme activators or inhibitors are looked for. This is because test reactions will finish either much sooner or much later when in the presence of potent activators or inhibitors, respectively.

In summary, harnessing the full power of high-throughput resources is now possible by adopting the 3-point kinetics assay as an alternative to traditional endpoint and kinetic-mode assays. A substantial decrease of false positive and false negative rates should be attained by improving the robustness and efficiency of the hit selection methodology. The following advantages of the 3-point method are emphasized:

- Simple implementation: 3-point monitoring can be implemented for a large number of screened reactions running simultaneously – none of the 3-points has to be located within the initial period of constant velocity. Moreover, the relative positions of the 3-points on the time-course may change from well-to-well or from plate-to-plate.

- Resilience to experimental variability: not only time-related variability but also experimental variability can be dealt with by the 3-point methodology. Random changes in, for example, substrate concentration, do not considerably affect the method resolution. Variations in concentration and/or activity of the enzyme can also be accounted for during hit selection.

- Interference-proof: the validation criteria provided by Eqs. 5.5, 5.6 and 5.7 are confirmed as powerful tools to eliminate artifacts. On the other hand, the proposed normalization method automatically corrects changes in background signal, which obviates the need of counter-assays to identify compound interferences.

- Admissible (and recommended) high substrate concentrations: Although desirable for improving the signal-to-background ratio and to detect uncompetitive inhibitors, high values of S_0 are traditionally associated to low sensitivity competitive inhibitors. With the 3-point method high substrate concentrations can be adopted without further constraints than the practical limits imposed by substrate solubility, substrate inhibition, and the desired overall duration of the assay.

- Fundamentally-based: both the validation and selection criteria are based on solid enzymology principles and not on subjective judgement.

- High sensitivity: the usage of LM coordinates warrants high sensitivity to minor kinetic variations, while the precursory application of robust validation criteria prevents false hit proliferation. Simple numerical examples show how the 3-point method has evident advantages in the detection of specific-type enzyme modifiers and of time-dependent modulation effects such as slow-binding inhibition.

As a general rule to HTS assay designers we recommend that the first and third points are located at beginning and conclusion of the assay, while the midpoint location should coincide with instantaneous substrate concentrations slightly above or below $\sim K_M$ depending on whether enzyme activators or inhibitors are looked for.

5.6. <u>Appendix</u>

5.6.1. Basic principles of the 3-point LM assay

For a general outline of the 3-point assay, consider the specific and catalytic enzyme modifier mechanisms (also known as competitive and uncompetitive mechanisms) presented in Figure

A5.1A. If quasi-equilibrium conditions are assumed for all binding steps [33, pp. 71-73], the following simplified definitions can be used to express the apparent constants K_m^{app} and V^{app} in terms of true equilibrium constants K_S and K_X and of the modifier concentration [X]:

$$K_m^{app} = K_s \frac{1 + \frac{[X]}{K_X}}{1 + \frac{[X]}{\alpha K_X}} \tag{A5.1a}$$

$$V^{app} = V \frac{1 + \beta\frac{[X]}{\alpha K_X}}{1 + \frac{[X]}{\alpha K_X}} \tag{A5.1b}$$

This means that the slope $(-K_m^{app})$ and vertical axis intercept (V^{app}) of the following LM straight lines are solely determined by the enzyme modifier concentration and by its mechanism of action:

$$\frac{\Delta P_i}{\Delta t_i} = V^{app} - K_m^{app}\left[-\frac{\ln\left(1 - \frac{\Delta P_i}{\Delta P_\infty}\right)}{\Delta t_i} \right] \tag{A5.2}$$

The proposed method of enzyme modifier detection relies on the measurement of product concentration P_i (or, alternatively, substrate concentration S_i) in three distinct moments (indexes $i = 0, 1, 2$). When the effects of the screening compounds on the calibration curve are not known beforehand, indirect monitoring of product concentration is performed using signal readout values F_i instead of P_i. Possible changes in background signal are normalized by using ΔF_i differences in the estimation of midpoint-centered ($i = 1$) LM coordinates:

$$X_1 = -\frac{\ln\left(1 - \frac{\Delta F_1}{\Delta F_2}\right)}{\Delta t_1} \tag{A5.3a}$$

$$Y_1 = \frac{\Delta F_1}{\Delta t_1} \tag{A5.3b}$$

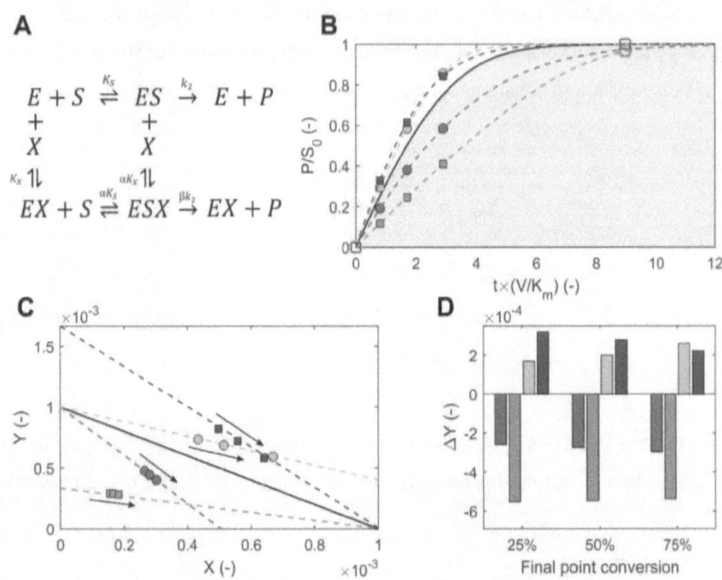

Figure A5.1. Possible outcomes of the 3-point assay expected for different enzyme modulation mechanisms and assuming different midpoint locations. (A) Simplified representation of the general modifier mechanism [34] (see Chapter 4, Sections 4.2.2 and 4.2.3), where K_S and K_X are dissociation constants, k_2 is the catalytic rate constant, and α and β are dimensionless constants used to differentiate specific ($\alpha > 1$) and catalytic ($\alpha < 1$) mechanisms of predominant enzyme inhibition ($\beta < 1$) and activation ($\beta > 1$). (B) Lines: progress curves representing S_0-normalized readouts for the control reaction (solid line, [X] = 0), and for test reactions (dashed lines, [X] = 0.1 µM) in the cases of specific activators ($\alpha = 0.2$, $\beta = 1$, light-blue), catalytic activators ($\alpha = 5$, $\beta = 5$, purple), specific inhibitors ($\alpha = +\infty$, $\beta = 0$, red), catalytic inhibitors ($\alpha = 0$, $\beta = 0$, green); further details concerning numerical simulation are given in Section 5.6.3.3 and Table A5.2. Full symbols: different midpoint locations. Open symbols: fixed initial and third point locations. Specific (circles) and catalytic (squares) mechanisms of either activation or inhibition are represented, respectively, outside or inside the area shaded in gray. (C) Lines: linearized progress curves expected from the combination of Eqs. A5.1 and A5.2 using the same model parameters and color code as in (B). Symbols: linearized midpoint coordinates corresponding to the 3-point selections made in (B) (Eq. A5.3). Arrows indicate progress in time. (D) Bars: influence of the midpoint location on the resolution of the 3-point method given as ΔY_1 differences (Eq. A5.4).

Positive hits are characterized by X_1 and Y_1 coordinates located significantly above (for activators) or below (for inhibitors) the boundary limit defined by the control LM curve (Figure A5.1C). During the screening of Atx-3 77Q effectors, the resolution of the method is computed in terms of the Y_1 difference:

$$\Delta Y_1 = Y_1 - (V - K_m X_1)\kappa \tag{A5.4}$$

where κ is the known proportionality factor converting product concentration units into F_i units in the absence of test compounds. Although no fixed procedure is imposed for the 3-point selection, narrow ranges of F_i variation are not recommended owing to the noise amplification originated by the logarithm term in Eq. A5.3a. Variations in the midpoint location have a slight effect on the resolution of the method, which, as expected, is principally influenced by the type and magnitude of the enzyme modifier mechanism in question (Figure A5.1D).

The first and third points do not have to coincide with the start and conclusion of complete catalytic reactions. Changing the initial condition to subsequent moments than $t_0 = 0$ has no major impact on the applicability of the method because the (X_1, Y_1) coordinates will continue to be located along the LM straight line (Figures A5.2A and A5.2B); this postponement of (t_0, P_0) may even prove beneficial, e.g., in enzyme modifier mechanisms of the specific type (Figure A5.2C).

On the contrary, if the value of F_2 is probed at earlier moments than the reaction completeness, the linearized coordinates will deviate from the linear trend and the resolution of the method will generally worsen, with possible occurrence of ΔY_1 sign change (Figure A5.3C). For this reason, the classification of hit compounds as activators or inhibitors should, in these cases, attend to the relative distribution of ΔY_1 values rather than to their positive or negative signal.

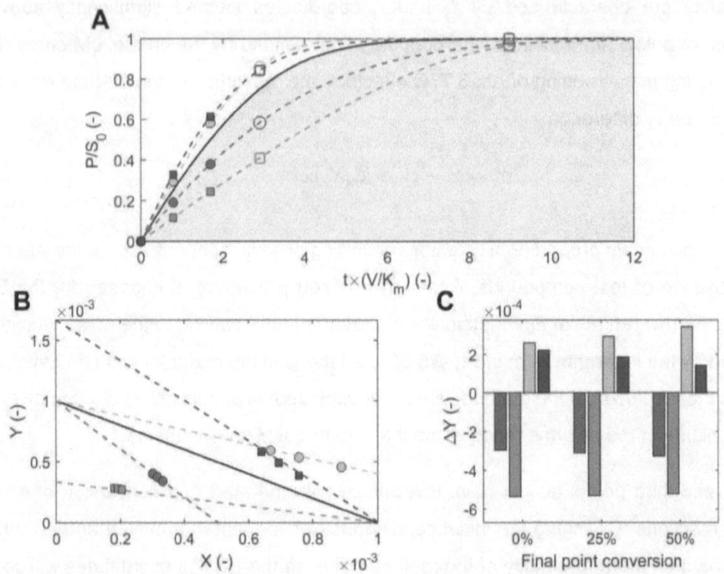

Figure A5.2. Possible outcomes of the 3-point assay for different initial point locations. (A) Lines: progress curves simulated as described in Figure A5.1. Full symbols: different initial point locations. Open symbols: fixed intermediate and third point locations. Specific (circles) and catalytic (squares) mechanisms of either activation or inhibition are represented, respectively, outside or inside the area shaded in gray. (B) Lines: linearized progress curves expected from the combination of Eqs. A5.1 and A5.2 using the same model parameters and color code as in (A). Symbols: linearized midpoint coordinates corresponding to the 3-point selections made in (A) (Eq. A5.3). (C) Bars: influence of the initial point location on the resolution of the 3-point method given as ΔY_1 differences (Eq. A5.4).

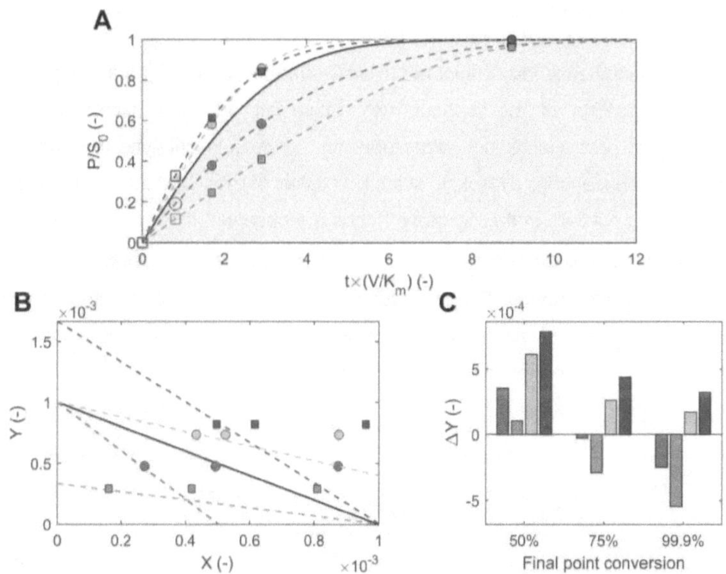

Figure A5.3. Possible outcomes of the 3-point assay for different third point locations. (A) Lines: progress curves simulated as described in Figure A5.1. Full symbols: different third point locations. Open symbols: fixed initial and intermediate point locations. Specific (circles) and catalytic (squares) mechanisms of either activation or inhibition are represented, respectively, outside or inside the area shaded in gray. (B) Lines: linearized progress curves expected from the combination of Eqs. A5.1 and A5.2 using the same model parameters and color code as in (A). Symbols: linearized midpoint coordinates corresponding to the 3-point selections made in (A) (Eq. A5.3). (C) Bars: influence of the third point location on the resolution of the 3-point method given as ΔY_1 differences (Eq. A5.4).

5.6.2. LM-Based validation criteria

Random experimental error may originate physically unrealistic (t_i, P_i) data. To give a simple example, the formation of air bubbles may cause an apparent decrease of product concentration with time that is not expectable for irreversible catalytic reactions. Subtler interferences can be identified by looking at the limit cases of linear and asymptotic exponential time-courses (Figure A5.4). In theory, these trends are observed if enzyme modifiers change the Michaelis constant to an apparent value of $K_m^{app} \approx 0$ (for linear curves) or $K_m^{app} \gg S_0$ (for asymptotic exponential curves). Progress curves in which product concentration increases faster than linear growth or slower than the slowest asymptotic exponential can be safely discarded owing to the occurrence of indeterminate spurious phenomena.

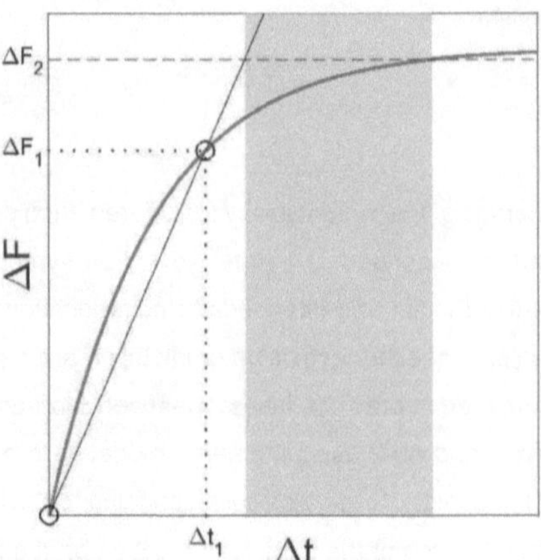

Figure A5.4. Linear (thin solid line) and asymptotic exponential curves (thick solid line) impose physical limits for the location of the 3-point coordinates. Once the two initial points are defined (symbols), the third concentration (dashed line) cannot be above the linear trend or below the asymptotic exponential trend. Shaded area: range of possible Δt_2 values.

Validation criteria can be defined for each set of three (t_i, F_i) coordinates by determining the values of Δt_2 that are admissible considering the time interval Δt_1 and the measured values of ΔF_1 and ΔF_2. Supra-linear trends are characterized by $\Delta F_2/\Delta F_1$ ratios exceeding $\Delta t_2/\Delta t_1$ (Figure A5.4). Therefore, the corresponding validation criterion can simply be expressed as:

$$\frac{\Delta t_1}{\Delta t_2} \leq \frac{\Delta F_1}{\Delta F_2} \tag{A5.5}$$

On the other hand, when the instantaneous substrate concentration is much lower than K_m^{app}, the asymptotic exponential is characterized by a time constant value of $\tau_\infty = K_m^{app}/V^{app}$ [20] (see Chapter 1, Section 1.4.1). The corresponding decay rate constant ($1/\tau_\infty$) can also be estimated from the midpoint coordinates assuming that the exponential decay of substrate concentration takes place since the beginning of the reaction:

$$\frac{1}{\tau_\infty} = \frac{-\ln(1 - \Delta F_1/\Delta F_\infty)}{\Delta t_1} \tag{A5.6}$$

This is in line with the expected for conditions of $K_m^{app} \gg S_0$. In this scenario, the lowest possible values of $\Delta F_2/\Delta t_2$ are defined by the instantaneous, third-point reaction rate $(dP/dt)_{t=t_2}$, which for steady-state conditions is given as $S_2 V^{app}/K_m^{app}$, or equivalently, as $(\Delta F_\infty - \Delta F_2)/\tau_\infty$. This result and Eq. A5.6 are used to establish the maximum $\Delta F_1/\Delta F_2$ ratio:

$$\frac{\Delta F_1}{\Delta F_2} \leq \frac{\Delta F_1}{\Delta F_\infty}\left(1 + \frac{\Delta t_1/\Delta t_2}{-\ln(1 - \Delta F_1/\Delta F_\infty)}\right) \tag{A5.7}$$

Finally, since $\Delta F_1/\Delta F_2$ cannot be greater than 1, maximum limits are established for $\Delta t_1/\Delta t_2$ in order to counteract $\Delta F_1/\Delta F_\infty$ values reaching close to the physical limit of 1. If $\Delta F_1/\Delta F_3$ is used as an overestimation of the (generally unknown) $\Delta F_1/\Delta F_\infty$ ratio, the last validation criterion is obtained by setting $\Delta F_1/\Delta F_2 = 1$ in Eq. A5.7:

$$\frac{\Delta t_1}{\Delta t_2} \leq -\ln\left(1 - \frac{\Delta F_1}{\Delta F_2}\right)\left(\frac{1}{\Delta F_1/\Delta F_2} - 1\right) \tag{A5.8}$$

As shown in Figure A5.5, this condition becomes more stringent than Eq. A5.5 for values of $\Delta F_1/\Delta F_2$ higher than ~ 0.6.

5.6.3. Numerical Methods

All numerical and analysis procedures were performed using Mathworks® MATLAB R2018a.

5.6.3.1. General data analysis procedures

To each progress curve of fluorescence increase over time, the baseline corresponding to the average of blank assays was subtracted. A moving average filter was then applied for data smoothing. A linear calibration curve was determined to convert fluorescence units into molar concentration units in the absence of test compounds.

5.6.3.2. Atx-7 77Q characterization

Time-courses were obtained for S_0 concentrations comprised between 0.09 and 1.50 µM and $E_0 = 0.25$ µM in quadruplicate. For each time-course, initial rate analysis was performed using the function *fitlm* for the determination of the instantaneous reaction rates (v_i) values using a linear regression over an initial interval of 3.5 hours. After determining the corresponding $S_i = S_0 - P_i$ values, kinetic parameters K_m and V were obtained by non-linear least-squares fitting of v_i vs S_i using the following equation:

$$v_i = V \frac{2S_i}{K_m + E_0 + S_i + \sqrt{(K_m + E_0 + S_i)^2 - 4E_0S_i}} \tag{A5.9}$$

whose applicability does not require a large enzyme excess over substrate [19]. Non-linear least-squares fitting and standard error determination (95% confidence level) was performed using the functions *fit* and *nlparci*, correspondingly.

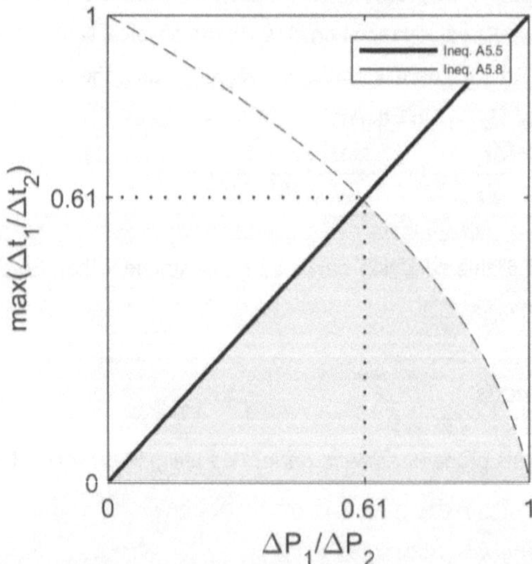

Figure A5.5. Maximum values of the $\Delta t_1/\Delta t_2$ ratio admitted by inequalities A5.5 (solid line) and A5.8 (dashed line).

5.6.3.3. Numerical simulations

The system of ordinary differential equations comprising Eqs. A5.10 describing the General Modifier Mechanism (GMM) [34] (Figure A5.1A) was numerically solved using the *ode15s* solver to describe the full reaction progress curves represented in Figure A5.1, Figure A5.2, Figure A5.3, and Figure 5.5 for the sets of simulation parameters summarized in Table A5.1 and Table A5.2.

$$\frac{d[S]}{dt} = k_{-1}[ES] - k_1[E][S] + k_{-5}[ESX] - k_5[EX][S] \tag{A5.10a}$$

$$\frac{d[P]}{dt} = k_2[ES] + k_6[ESX] \tag{A5.10b}$$

$$\frac{d[E]}{dt} = k_{-1}[ES] - k_1[E][S] + k_2[ES] - k_3[E][X] + k_{-3}[EX] \tag{A5.10c}$$

$$\frac{d[ES]}{dt} = k_1[E][S] - k_{-1}[ES] - k_2[ES] - k_4[ES][X] + k_{-4}[ESX] \tag{A5.10d}$$

$$\frac{d[EX]}{dt} = k_3[E][X] - k_{-3}[EX] - k_5[EX][S] + k_{-5}[ESX] \tag{A5.10e}$$

$$\frac{d[ESX]}{dt} = k_4[ES][X] - k_{-4}[ESX] + k_5[EX][S] - k_{-5}[ESX] - k_6[ESX] \tag{A5.10f}$$

$$\frac{d[X]}{dt} = k_{-3}[EX] - k_3[E][X] + k_{-4}[ESX] - k_4[ES][X] \tag{A5.10g}$$

where $[E]$, $[S]$, and $[X]$ are the concentration values of total enzyme, substrate, and present modifier, and $[ES]$, $[EX]$, and $[ESX]$ are the concentration values of the complexes formed by enzyme, substrate and modifier. The indicated rate constants are related to the dissociation constants present in the GMM reaction scheme (Figure A5.1A) by the following relations: $K_s = k_{-1}/k_1$ (substrate dissociation constant), $K_3 = k_{-3}/k_3 = K_X$ (dissociation constant of the specific component), $K_4 = k_{-4}/k_4 = \alpha K_X$ (dissociation constant of the catalytic component), $K_5 = k_{-5}/k_5 = \alpha K_s$. Parameters α and β are, respectively, the reciprocal allosteric coupling constant between modifier and substrate, and the factor by which the modifying compound alters the catalytic constant k_2.

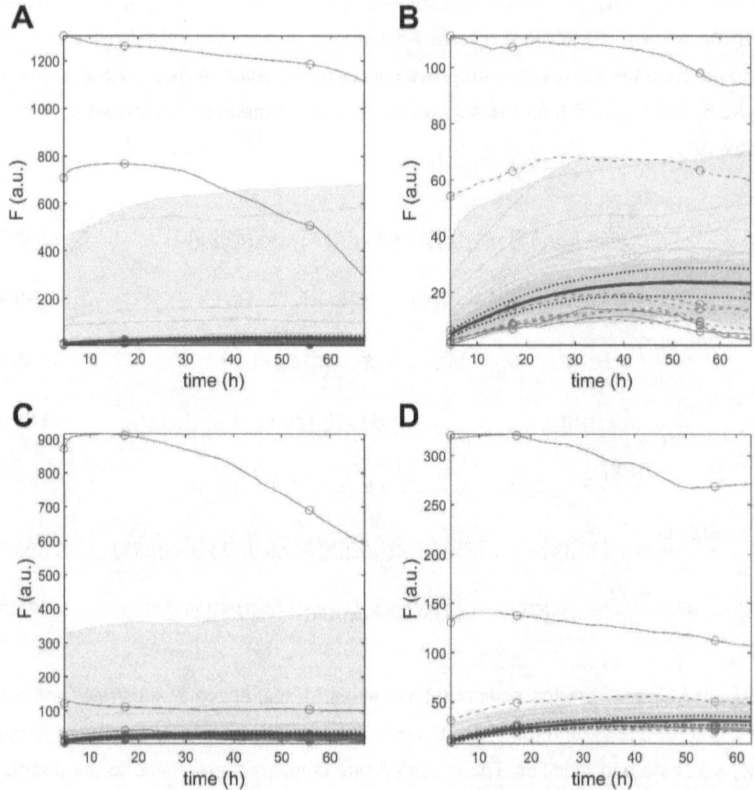

Figure A5.6. Full scale graph of Figure 5.2. (A-D) Progress curves showing mean and standard deviation values of control reaction readouts (solid and dashed blue lines, respectively); excluded (red lines) and validated (gray lines) screening reactions; and the 3-point positions (symbols) along the excluded progress curves. Filters to eliminate spurious results are successively applied based on Eqs. 5.5 (solid red lines), 5.6 (no results excluded) and 5.7 (dashed red lines). Different panels present the results obtained in microplates (A) #1, (B) #2, (C) #3, (D) #4. Areas shaded in light gray represent intervals where validated test reactions can be found.

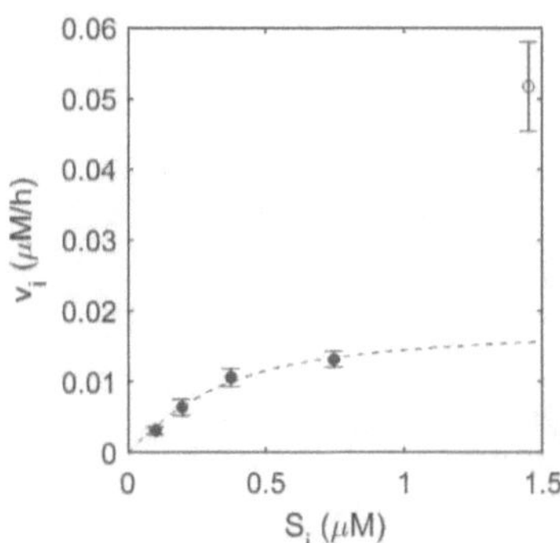

Figure A5.7. Kinetic characterization of Atx-3 77Q. Symbols and error bars: means and standard deviations of the instantaneous reaction rates (v_i) represented as a function of S_i. Dashed line: numerical adjustment of Eq. A5.9 using selected experimental data (closed symbols). Excluded reaction rates (open symbols) are obtained for substrate concentrations (>1 µM) that are not recommended by the manufacturer. Fitted results: $K_m = 0.191 \pm 0.070$ µM, $V = 0.018 \pm 0.002$ µM/h, $R^2 = 0.991$.

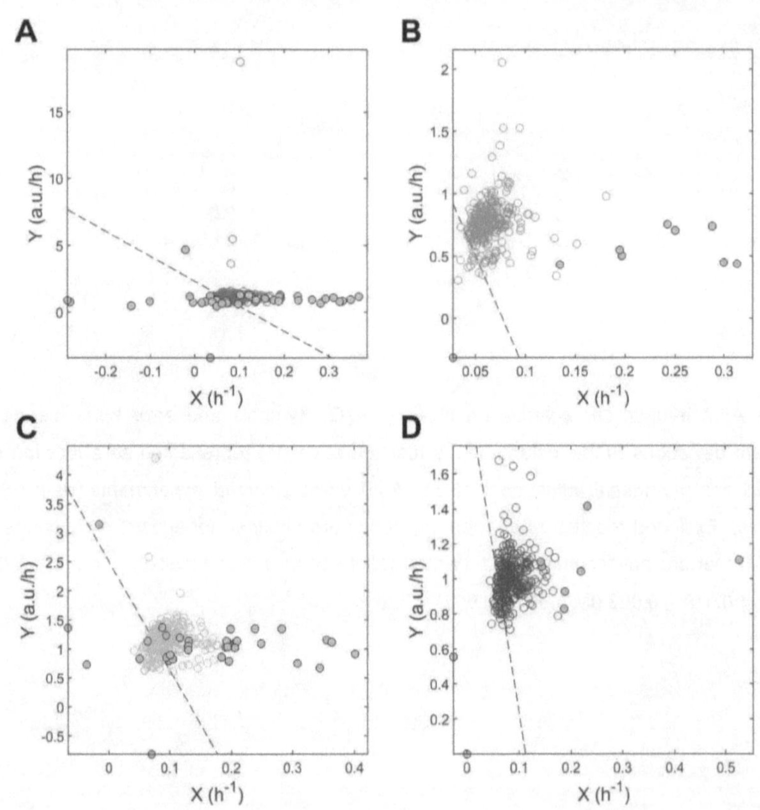

Figure A5.8. Full scale graph of Figure 5.3 showing LM representation of screening reactions. (A-D) Symbols: intermediate LM coordinates computed for validated (open symbols) and excluded (closed symbols in gray) results using Eq. 5.3; for graphical representation purposes, argument of the logarithm in Eq. 5.3a is determined as its module. Dashed Lines: theoretical LM straight lines computed from Eq. 5.2 using the values of K_m and V obtained during Atx-3 77Q assay characterization (Figure A5.7); parameter V is corrected by a factor of (A) 0.94, (B) 0.59, (C) 1.02, (D) 0.86 accounting for the different Atx-3 77Q activities measured at 0.375 μM Ub-AMC during assay characterization and during the control reactions in each microplate.

5.6.5. Tables

Table A5.1. Reaction parameters used for simulation of full progress curves under the GMM and a case of slow-binding inhibition (Eqs. A5.10) presented in Figure 5.5, obtained in the presence of $E_0 = 0.01$ µM and $S_0 = 10$ µM, for control reaction parameters $K_m = 1$ µM and $V = 10^{-3}$ µM/s and a time interval of $t = [0, 1.2 \times 10^4]$ s.

Kinetic Scenario	k_1 ($\mu M^{-1}s^{-1}$)	k_{-1} (s^{-1})	k_2 (s^{-1})	k_3 ($\mu M^{-1}s^{-1}$)	k_{-3} (s^{-1})	k_4 ($\mu M^{-1}s^{-1}$)	k_{-4} (s^{-1})	k_5 ($\mu M^{-1}s^{-1}$)	k_{-5} (s^{-1})	k_6 (s^{-1})	α (–)	β (–)
Control				0	0	0	0	0	0	0	-	-
Linear Specific Inhibition				40	1	0	10^{10}	0	10^{10}	0	$+\infty$	0
Linear Catalytic Inhibition	10	9.9	0.1	0	10^{10}	20	2	10^{10}	0	0	0	0
Slow-binding Inhibition				5×10^{-2}	5×10^{-5}	0	10^{10}	0	10^{10}	0	-	-

Table A5.2. Reaction parameters used for simulation of full progress curves under the GMM (Eqs. A5.10) presented in Figure A5.1, Figure A5.2, and Figure A5.3, obtained in the presence of $E_0 = 0.01$ µM, $S_0 = 2$ µM, and $[X] = 0.1$ µM, for control reaction parameters $K_m = 1$ µM and $V = 10^{-3}$ µM/s, and a time interval of $t = [0, 1.2 \times 10^4]$ s.

GMM scenario	k_1 ($\mu M^{-1}s^{-1}$)	k_{-1} (s^{-1})	k_2 (s^{-1})	k_3 ($\mu M^{-1}s^{-1}$)	k_{-3} (s^{-1})	k_4 ($\mu M^{-1}s^{-1}$)	k_{-4} (s^{-1})	k_5 ($\mu M^{-1}s^{-1}$)	k_{-5} (s^{-1})	k_6 (s^{-1})	α (–)	β (–)
Control				0	0	0	0	0	0	0	-	-
Linear Specific Inhibition				10	1	0	10^{10}	0	10^{10}	0	$+\infty$	0
Linear Catalytic Inhibition	10	9.9	0.1	0	10^{10}	20	1	10^{10}	0	0	0	0
Hyperbolic Specific Activation				10	5	10	1	100	20	0.1	0.2	1
Hyperbolic Catalytic Activation				10	1	10	5	10	50	0.5	5	5

References

1. Michaelis, L. and Menten, M. (1913), *Die Kinetik der Invertinwirkung.* Biochemische Zeitschrift. 49:333–369.

2. Copeland, R.A., (2013), *Evaluation of enzyme inhibitors in drug discovery: a guide for medicinal chemists and pharmacologists.* John Wiley & Sons.

3. Copeland, R.A. (2003), *Mechanistic considerations in high-throughput screening.* Analytical Biochemistry. 320(1):1-12.

4.   Gutiérrez, O.A. and Danielson, U.H. (2006), *Detection of competitive enzyme inhibition with end point progress curve data.* Analytical Biochemistry. 358(1):11-19.

5.   Inglese, J., *et al.* (2007), *High-throughput screening assays for the identification of chemical probes.* Nature Chemical Biology. 3(8):466.

6.   Orsi, B.A. and Tipton, K.F. (1979), *Kinetic analysis of progress curves.* Methods in Enzymology. 63:159-83.

7.   Tholander, F. (2012), *Improved inhibitor screening experiments by comparative analysis of simulated enzyme progress curves.* PloS one. 7(10):e46764.

8.   Gunawardena, J. (2012), *Some lessons about models from Michaelis and Menten.* Molecular Biology of the Cell. 23(4):517-519.

9.   Pinto, M.F., *et al.* (2019), *A simple linearization method unveils hidden enzymatic assay interferences.* Biophysical Chemistry. 252:106193.

10.   Wu, G., Yuan, Y., and Hodge, C.N. (2003), *Determining appropriate substrate conversion for enzymatic assays in high-throughput screening.* Journal of Biomolecular Screening. 8(6):694-700.

11.   Sárkány, Z., *et al.* (2019), *Chemical Kinetic Strategies for High‐Throughput Screening of Protein Aggregation Modulators.* Chemistry–An Asian Journal. 14(4):500-508.

12.   Ma, H., *et al.* (2005), *Nanoliter homogenous ultra-high throughput screening microarray for lead discoveries and IC50 profiling.* Assay and Drug Development Technologies. 3(2):177-187.

13.   Petersen, K.J., *et al.* (2014), *Fluorescence lifetime plate reader: Resolution and precision meet high-throughput.* Review of Scientific Instruments. 85(11):113101.

14.   Bisswanger, H. (2014), *Enzyme assays.* Perspectives in Science. 1(1–6):41-55.

15.   Cornish-Bowden, A. (1975), *The use of the direct linear plot for determining initial velocities.* Biochemical Journal. 149(2):305-312.

16.   Gribbon, P. and Sewing, A. (2003), *Fluorescence readouts in HTS: no gain without pain?* Drug Discovery Today. 8(22):1035-1043.

17.   Shoichet, B.K. (2006), *Screening in a spirit haunted world.* Drug Discovery Today. 11(13-14):607-615.

18.   Simeonov, A. and Davis, M.I., (2018), *Interference with fluorescence and absorbance.* Assay Guidance Manual [Internet]. Eli Lilly & Company and the National Center for Advancing Translational Sciences.

19.   Pinto, M.F., *et al.* (2015), *Enzyme kinetics: the whole picture reveals hidden meanings.* FEBS Journal. 282(12):2309-2316.

20.   Pinto, M.F. and Martins, P.M. (2016), *In search of lost time constants and of non-Michaelis-Menten parameters.* Perspectives in Science. 9:8-16.

21. Silva, A., de Almeida, A.V., and Macedo-Ribeiro, S. (2018), *Polyglutamine expansion diseases: More than simple repeats.* Journal of Structural Biology. 201(2):139-154.

22. Almeida, B., *et al.* (2015), *SUMOylation of the brain-predominant Ataxin-3 isoform modulates its interaction with p97.* Biochimica et Biophysica Acta (BBA) - Molecular Basis of Disease. 1852(9):1950-1959.

23. Gales, L., *et al.* (2005), *Towards a Structural Understanding of the Fibrillization Pathway in Machado-Joseph's Disease: Trapping Early Oligomers of Non-expanded Ataxin-3.* Journal of Molecular Biology. 353(3):642-654.

24. Silva, A., *et al.* (2017), *Distribution of Amyloid-Like and Oligomeric Species from Protein Aggregation Kinetics.* Angewandte Chemie International Edition. 56(45):14042-14045.

25. Burnett, B., Li, F., and Pittman, R.N. (2003), *The polyglutamine neurodegenerative protein ataxin-3 binds polyubiquitylated proteins and has ubiquitin protease activity.* Human Molecular Genetics. 12(23):3195-3205.

26. Gribbon, P., *et al.* (2005), *Evaluating real-life high-throughput screening data.* Journal of Biomolecular Screening. 10(2):99-107.

27. Zhang, J.-H., Chung, T.D., and Oldenburg, K.R. (1999), *A simple statistical parameter for use in evaluation and validation of high throughput screening assays.* Journal of biomolecular screening. 4(2):67-73.

28. Tirat, A., *et al.* (2005), *Synthesis and characterization of fluorescent ubiquitin derivatives as highly sensitive substrates for the deubiquitinating enzymes UCH-L3 and USP-2.* Analytical Biochemistry. 343(2):244-255.

29. Hanley, Q.S. (2019), *the Distribution of standard Deviations Applied to High throughput screening.* Scientific Reports. 9(1):1268.

30. Yang, J., Copeland, R.A., and Lai, Z. (2009), *Defining balanced conditions for inhibitor screening assays that target bisubstrate enzymes.* Journal of Biomolecular Screening. 14(2):111-20.

31. Murie, C., *et al.* (2015), *Improving detection of rare biological events in high-throughput screens.* Journal of Biomolecular Screening. 20(2):230-241.

32. Acker, M.G. and Auld, D.S. (2014), *Considerations for the design and reporting of enzyme assays in high-throughput screening applications.* Perspectives in Science. 1(1–6):56-73.

33. Baici, A., (2015), *Kinetics of Enzyme-Modifier Interactions,* 1st ed. Springer.

34. Botts, J. and Morales, M. (1953), *Analytical description of the effects of modifiers and of enzyme multivalency upon the steady state catalyzed reaction rate.* Transactions of the Faraday Society. 49(0):696-707.

Conclusions & Future Work

The general and most widespread formalism for enzyme kinetics analysis remains that proposed by Michaelis and Menten in 1913. These authors derived a simplified equation describing the dependency of initial reaction rates on initial substrate concentration valid for the single active site, single substrate reaction mechanism later described by Briggs and Haldane in 1925. The application of the MM equation requires conditions of enzyme and substrate concentration that may not be realistic under cell-like environments. Answering these limitations, the PEA model was proposed in 2015 as the unconstrained closed-form solution of the (non-inhibited) single active-site enzymatic mechanism. The PEA model paved the way to a series of new and robust kinetic tools that constitute the main achievements of this Doctoral book.

Re-inspection of Michaelis and Menten's original work allowed the dissection of unexplored elements of their analysis and led to the proposal of the $(EA)^2$ assay to characterize enzyme activity, efficiency, and affinity from single progress curves. This methodology is particularly important in view of upcoming studies addressing biologically realistic conditions how they compare with *in vitro* reaction settings.

The adequate application of kinetic methodologies such as the PEA Model or the $(EA)^2$ assay required an experimental setup devoid of assay interferences. Hence, our next step involved the proposal of the LM to assess and validate enzymatic assays. This new tool is very sensitive to alterations in the measured time-courses, which permits strict selection of valid kinetic data for kinetic parameter estimation. It is expected that the LM will continue to contribute for the accuracy and reproducibility of enzymology data by means of rigorous quality-control implementation even when assay interferences are not suspected beforehand and without the need of additional experiments.

The LM was then further expanded to automate the detection of assay interferences and estimate unbiased kinetic parameters for successfully validated assays. As a major achievement of this PhD project, the webserver "interferENZY" was successfully created and is now available to any user interested in validating kinetic data obtained from continuous and end-point assays. Kinetic parameter determination is also possible using the webserver with the significant advantage of not relying on erratic initial rate measurements. The interferENZY webserver answers the current demand for transparent and standardized methodologies in enzyme kinetics that could lead to enhanced reproducibility in experimental data reporting. The following future developments are envisaged for interferENZY:

(i)   Kinetic characterization of enzyme modifiers. This will be implemented through the direct application of the GMM. The user will be able to upload a number of datasets obtained in the presence of different modifier concentrations. The underlying kinetic mechanism will be characterized based on the estimated values of the apparent kinetic parameters. This is possible because the LM equation on which this interferENZY is based is compatible with all 17 basic mechanisms documented for the GMM.

(ii)   Analysis of HTS data for hit detection and seriation. This will be implemented by using an automatic algorithm based on the 3-point kinetic assay procedures (including outlier exclusion principles), which will be compatible with datasets containing more than 3 points per curve.

Chemical kinetic analysis in the presence of enzyme modulators was illustrated during the characterization of the potential inhibitory activity of the bacterial ortholog of frataxin, CyaY, on the enzymatic activity of the IscS-IscU desulfurase-scaffold system. This model enzymatic system is of interest to the study of the pathophysiology of the neurodegenerative disease FRDA, where frataxin deficiency has a central role. Through the classic GMM analysis originally described by Botts and Morales in 1953, it was possible to characterize CyaY as a putative hyperbolic catalytic inhibitor. In the near future, complementary experiments will be performed in order to expand this initial body of evidence.

The 3-point kinetic assay was also developed in view of systematic analysis of HTS experiments for enhanced hit compound detection. A library of ~1200 FDA-approved molecules was screened for possible modulators of the deubiquinating activity of a pathogenic variant of ataxin-3, an aggregation-prone protein participating in the pathophysiology of the neurodegenerative Machado-Joseph Disease [1-3]. Ataxin-3 aggregation is thought to lead to loss of enzymatic activity, thereby affecting the functions of protein quality control and of protein homeostasis maintenance associated to this polyQ protein [3-5].

Hit and artifact detection rates under the 3-point method were considerable higher than by adopting traditional initial rate analysis, thus demonstrating how the efficiency and accuracy of primary screenings can be improved in future drug discovery campaigns. We are enthusiastic about preliminary dose response results confirming that the deubiquinating activity of ataxin-3 can be rescued by drug-like compounds. Certainly, the selected hit compounds will continue to be studied in the near future following drug repurposing strategies that have been proved successful in the past [6,7].

The present Doctoral book shows the importance of proper enzymatic assay optimization and setup, together with suitable biophysical modelization, for full and accurate description of enzyme kinetics. On a fundamental level, this assures the validity of documented kinetic parameters and strengthens possible inferences concerning catalytic mechanisms. Only with this degree of control will be possible to tackle the challenges posed by cell-like enzymatic experimentation and to explore the critical region of biological conditions predicted

Model. On a practical level, this post-Michaelis-Menten framework brings forward new and more robust kinetic tools expected to increase the sensitivity and accuracy of drug-like compounds targeting enzymes of interest in pharmaceutical research.

References

1. Matos, C.A., de Macedo-Ribeiro, S., and Carvalho, A.L. (2011), *Polyglutamine diseases: the special case of ataxin-3 and Machado-Joseph disease.* Progress in Neurobiology. 95(1):26-48.

2. Scheel, H., Tomiuk, S., and Hofmann, K. (2003), *Elucidation of ataxin-3 and ataxin-7 function by integrative bioinformatics.* Human Molecular Genetics. 12(21):2845-2852.

3. Burnett, B., Li, F., and Pittman, R.N. (2003), *The polyglutamine neurodegenerative protein ataxin-3 binds polyubiquitylated proteins and has ubiquitin protease activity.* Human Molecular Genetics. 12(23):3195-3205.

4. Doss-Pepe, E.W., *et al.* (2003), *Ataxin-3 Interactions with Rad23 and Valosin-Containing Protein and Its Associations with Ubiquitin Chains and the Proteasome Are Consistent with a Role in Ubiquitin-Mediated Proteolysis.* Molecular and Cellular Biology. 23(18):6469-6483.

5. Zhong, X. and Pittman, R.N. (2006), *Ataxin-3 binds VCP/p97 and regulates retrotranslocation of ERAD substrates.* Human Molecular Genetics. 15(16):2409-2420.

6. Roy, A. (2018), *Early probe and drug discovery in academia: a minireview.* High-throughput. 7(1):4.

7. Sant'Anna, R., *et al.* (2016), *Repositioning tolcapone as a potent inhibitor of transthyretin amyloidogenesis and associated cellular toxicity.* Nature Communications. 7:10787.

Full Reference List

Acker, M.G. and Auld, D.S. (2014), *Considerations for the design and reporting of enzyme assays in high-throughput screening applications.* Perspectives in Science. 1(1–6):56-73.

Adinolfi, S., et al. (2002), *A structural approach to understanding the iron-binding properties of phylogenetically different frataxins.* Human Molecular Genetics. 11(16):1865-77.

Adinolfi, S., et al. (2009), *Bacterial frataxin CyaY is the gatekeeper of iron-sulfur cluster formation catalyzed by IscS.* Nature Structural & Molecular Biology. 16(4):390-6.

Adinolfi, S., et al. (2018), *The molecular bases of the dual regulation of bacterial iron sulfur cluster biogenesis by CyaY and IscX.* Frontiers in Molecular Biosciences. 4:97.

Aldrich, C., et al. (2017), *The Ecstasy and Agony of Assay Interference Compounds.* ACS Central Science. 3(3):143-147.

Almeida, B., et al. (2015), *SUMOylation of the brain-predominant Ataxin-3 isoform modulates its interaction with p97.* Biochimica et Biophysica Acta (BBA) - Molecular Basis of Disease. 1852(9):1950-1959.

Baici, A. (1981), *The Specific Velocity Plot.* European Journal of Biochemistry. 119(1):9-14.

Baici, A., (2015), *Kinetics of Enzyme-Modifier Interactions,* 1st ed. Springer.

Bajzer, Z. and Strehler, E.E. (2012), *About and beyond the Henri-Michaelis-Menten rate equation for single-substrate enzyme kinetics.* Biochemical and Biophysical Research Communications. 417(3):982-985.

Barshop, B.A., Wrenn, R.F., and Frieden, C. (1983), *Analysis of numerical methods for computer simulation of kinetic processes: development of KINSIM--a flexible, portable system.* Analytical biochemistry. 130(1):134-145.

Bäuerle, F., Zotter, A., and Schreiber, G. (2016), *Direct determination of enzyme kinetic parameters from single reactions using a new progress curve analysis tool.* Protein Engineering, Design and Selection. 30(3):151-158.

Ben Halima, S., et al. (2016), *Specific Inhibition of β-Secretase Processing of the Alzheimer Disease Amyloid Precursor Protein.* Cell Reports. 14(9):2127-2141.

Berberan-Santos, M.N. (2010), *A General Treatment of Henri-Michaelis-Menten Enzyme Kinetics: Exact Series Solution and Approximate Analytical Solutions.* MATCH Communications in Mathematical and in Computer Chemistry. 63:283-318.

Bersani, A.M. and Dell'Acqua, G. (2011), *Asymptotic expansions in enzyme reactions with high enzyme concentrations.* Mathematical Methods in the Applied Sciences. 34(16):1954-1960.

Bevc, S., et al. (2011), *ENZO: a web tool for derivation and evaluation of kinetic models of enzyme catalyzed reactions.* PloS one. 6(7):e22265-e22265.

Bisswanger, H. (2014), *Enzyme assays.* Perspectives in Science. 1(1–6):41-55.

Boatright, K.M. and Salvesen, G.S. (2003), *Mechanisms of caspase activation*. Current Opinion in Cell Biology. 15(6):725-731.

Bose, K., *et al.* (2003), *An Uncleavable Procaspase-3 Mutant Has a Lower Catalytic Efficiency but an Active Site Similar to That of Mature Caspase-3*. Biochemistry. 42(42):12298-12310.

Botts, J. and Morales, M. (1953), *Analytical description of the effects of modifiers and of enzyme multivalency upon the steady state catalyzed reaction rate*. Transactions of the Faraday Society. 49(0):696-707.

Boucher, D., Duclos, C., and Denault, J.-B., (2014), Caspases, Paracaspases and Metacaspases: Methods and Protocols, ed. P.V. Bozhkov and G. Salvesen. Humana Press, Chp.

Bridwell-Rabb, J., *et al.* (2012), *Effector Role Reversal during Evolution: The Case of Frataxin in Fe–S Cluster Biosynthesis*. Biochemistry. 51(12):2506-2514.

Briggs, G.E. and Haldane, J.B.S. (1925), *A note on the kinetics of enzyme action*. Biochemical Journal. 19:338-339.

Brown, A.J. (1902), *XXXVI.—Enzyme action*. Journal of the Chemical Society, Transactions. 81:373-388.

Buchholz, P.C.F., *et al.* (2016), *BioCatNet: A Database System for the Integration of Enzyme Sequences and Biocatalytic Experiments*. ChemBioChem. 17(21):2093-2098.

Burnett, B., Li, F., and Pittman, R.N. (2003), *The polyglutamine neurodegenerative protein ataxin-3 binds polyubiquitylated proteins and has ubiquitin protease activity*. Human Molecular Genetics. 12(23):3195-3205.

Campuzano, V., *et al.* (1996), *Friedreich's ataxia: autosomal recessive disease caused by an intronic GAA triplet repeat expansion*. Science. 271(5254):1423-7.

Campuzano, V., *et al.* (1997), *Frataxin is reduced in Friedreich ataxia patients and is associated with mitochondrial membranes*. Human Molecular Genetics. 6(11):1771-80.

Cao, W. and De La Cruz, E.M. (2013), *Quantitative full time course analysis of nonlinear enzyme cycling kinetics*. Scientific Reports. 3:2658.

Cárdenas, M.L. and Cornish-Bowden, A. (1989), *Characteristics necessary for an interconvertible enzyme cascade to generate a highly sensitive response to an effector*. Biochemical Journal. 257(2):339.

Chaplin, M.F. and Bucke, C., (1990), *Enzyme technology*. CUP Archive, pp.32.

Choi, B., Rempala, G.A., and Kim, J.K. (2017), *Beyond the Michaelis-Menten equation: Accurate and efficient estimation of enzyme kinetic parameters*. Scientific Reports. 7(1):17018.

Cook, A.G., Tolliver, R.M., and Williams, J.E. (1994), *The Blue Bottle Experiment Revisited: How Blue? How Sweet?* Journal of Chemical Education. 71(2):160.

Copeland, R.A. (2003), *Mechanistic considerations in high-throughput screening*. Analytical Biochemistry. 320(1):1-12.

Copeland, R.A., (2013), *Evaluation of enzyme inhibitors in drug discovery: a guide for medicinal chemists and pharmacologists*. John Wiley & Sons.

Cornish-Bowden, A. (1975), *The use of the direct linear plot for determining initial velocities*. Biochemical Journal. 149(2):305-312.

Cornish-Bowden, A. (1987), *The Time Dimension in Steady-state Kinetics: A Simplified Representation of Control Coefficients*. Biochemical Education. 15(3):144-146.

Cornish-Bowden, A. (2013), *The origins of enzyme kinetics*. FEBS Letters. 587:2725-2730.

Cornish-Bowden, A. (2015), *One hundred years of Michaelis–Menten kinetics*. Perspectives in Science. 4:3-9.

Cornish-Bowden, A., (2012), *Fundamentals of Enzyme Kinetics*, 4th ed. Wiley-Blackwell (Weinheim, Germany).

Corral-Rodríguez, M.Á., *et al.* (2009), *Tick-derived Kunitz-type inhibitors as antihemostatic factors*. Insect Biochemistry and Molecular Biology. 39(9):579-595.

Corral-Rodríguez, M.Á., *et al.* (2010), *Leech-Derived Thrombin Inhibitors: From Structures to Mechanisms to Clinical Applications*. Journal of Medicinal Chemistry. 53(10):3847-3861.

Crawley, J.T.B., *et al.* (2007), *The central role of thrombin in hemostasis*. Journal of Thrombosis and Haemostasis. 5(s1):95-101.

Cupp-Vickery, J.R., Urbina, H., and Vickery, L.E. (2003), *Crystal Structure of IscS, a Cysteine Desulfurase from Escherichia coli*. Journal of Molecular Biology. 330(5):1049-1059.

di Maio, D., *et al.* (2017), *Understanding the role of dynamics in the iron sulfur cluster molecular machine*. Biochimica et Biophysica Acta (BBA) - General Subjects. 1861(1, Part A):3154-3163.

Dormand, J.R. and Prince, P.J. (1980), *A family of embedded Runge-Kutta formulae"*. Journal of Computational and Applied Mathematics. 6:19-26.

Doss-Pepe, E.W., *et al.* (2003), *Ataxin-3 Interactions with Rad23 and Valosin-Containing Protein and Its Associations with Ubiquitin Chains and the Proteasome Are Consistent with a Role in Ubiquitin-Mediated Proteolysis*. Molecular and Cellular Biology. 23(18):6469-6483.

Duggleby, R.G. (2001), *Quantitative analysis of the time courses of enzyme-catalyzed reactions*. Methods. 24(2):168-174.

Eicher, J., Snoep, J., and Rohwer, J. (2012), *Determining Enzyme Kinetics for Systems Biology with Nuclear Magnetic Resonance Spectroscopy*. Metabolites. 2(4):818.

Eisenthal, R., Danson, M.J., and Hough, D.W. (2007), *Catalytic efficiency and k_{cat}/K_M: a useful comparator?* TRENDS in Biotechnology. 25(6):247-249.

Feng, B.Y. and Shoichet, B.K. (2006), *A detergent-based assay for the detection of promiscuous inhibitors.* Nat Protoc. 1(2):550-3.

Ferreira, C., *et al.* (2019), *Protein crystals as a key for deciphering macromolecular crowding effects on biological reactions.* Submitted.

Fersht, A., (1999), *Structure and Mechanism in Protein Science*, 2nd ed. W.H. Freeman and Company (New York).

Finn, N. and Kemp, M., (2014), Systems Biology Approaches to Enzyme Kinetics: Analyzing Network Models of Drug Metabolism. Methods in Molecular Biology. Vol. 1113. Humana Press, Chp. 15.

Fleisher, G.A. (1953), *Curve Fitting of Enzymatic Reactions Based on the Michaelis—Menten Equation[1].* Journal of the American Chemical Society. 75(18):4487-4490.

Foury, F. and Cazzalini, O. (1997), *Deletion of the yeast homologue of the human gene associated with Friedreich's ataxia elicits iron accumulation in mitochondria.* FEBS Letters. 411(2-3):373-7.

Francischetti, I.M., *et al.* (2009), *The role of saliva in tick feeding.* Front Biosci (Landmark Ed). 14:2051-88.

Gales, L., *et al.* (2005), *Towards a Structural Understanding of the Fibrillization Pathway in Machado-Joseph's Disease: Trapping Early Oligomers of Non-expanded Ataxin-3.* Journal of Molecular Biology. 353(3):642-654.

Gerber, J., Muhlenhoff, U., and Lill, R. (2003), *An interaction between frataxin and Isu1/Nfs1 that is crucial for Fe/S cluster synthesis on Isu1.* EMBO Reports. 4(9):906-11.

Gielen, F., *et al.* (2013), *A Fully Unsupervised Compartment-on-Demand Platform for Precise Nanoliter Assays of Time-Dependent Steady-State Enzyme Kinetics and Inhibition.* Analytical Chemistry. 85(9):4761-4769.

Gloria, P.M.C., *et al.* (2011), *Aspartic vinyl sulfones: inhibitors of a caspase-3-dependent pathway.* Eur J Med Chem. 46(6):2141-6.

Gribbon, P. and Sewing, A. (2003), *Fluorescence readouts in HTS: no gain without pain?* Drug Discovery Today. 8(22):1035-1043.

Gribbon, P., *et al.* (2005), *Evaluating real-life high-throughput screening data.* Journal of Biomolecular Screening. 10(2):99-107.

Grosch, J.-H., *et al.* (2017), *Influence of the experimental setup on the determination of enzyme kinetic parameters.* Biotechnology Progress. 33(1):87-95.

Gunawardena, J. (2012), *Some lessons about models from Michaelis and Menten.* Molecular Biology of the Cell. 23(4):517-519.

Gutiérrez, O.A. and Danielson, U.H. (2006), *Detection of competitive enzyme inhibition with end point progress curve data.* Analytical Biochemistry. 358(1):11-19.

Hanley, Q.S. (2019), *the Distribution of standard Deviations Applied to High throughput screening.* Scientific Reports. 9(1):1268.

Hanson, S.M. and Schnell, S. (2008), *Reactant Stationary Approximation in Enzyme Kinetics.* The Journal of Physical Chemistry A. 112:8654-8658.

Hanson, S.M. and Schnell, S. (2008), *Reactant Stationary Approximation in Enzyme Kinetics.* The Journal of Physical Chemistry A. 112(37):8654-8658.

He, Y., *et al.* (2004), *Yeast frataxin solution structure, iron binding, and ferrochelatase interaction.* Biochemistry. 43(51):16254-62.

Henderson, P.J.F. (1973), *Steady-state enzyme kinetics with high-affinity substrates or inhibitors. A statistical treatment of dose–response curves.* Biochemical Journal. 135(1):101.

Henri, V. (1902), *Théorie générale de l'action des quelques diastases.* Comptes rendus hebdomadaires des séances de l'Académie des sciences. 135:916-919.

Henri, V., (1903), *Lois Générales de l'action des Diastases.* Librairie Scientifique A. Hermann (Paris).

Holdgate, G.A., Meek, T.D., and Grimley, R.L. (2017), *Mechanistic enzymology in drug discovery: a fresh perspective.* Nature Reviews: Drug Discovery. 17:115.

Inglese, J., *et al.* (2007), *High-throughput screening assays for the identification of chemical probes.* Nature Chemical Biology. 3(8):466.

Johnson, K.A. and Goody, R.S. (2011), *The Original Michaelis Constant: Translation of the 1913 Michaelis-Menten Paper.* Biochemistry. 50:8264-8269.

Johnson, K.A., Simpson, Z.B., and Blom, T. (2009), *Global Kinetic Explorer: A new computer program for dynamic simulation and fitting of kinetic data.* Analytical Biochemistry. 387(1):20-29.

Kendall, M.G. (1938), *A new measure of rank correlation.* Biometrika. 30(1-2):81-93.

Kettner, C. and Hicks, M.G. (2015), *Celebrating the 100th anniversary of Michaelis Menten-Kinetics.* Perspectives in Science. 4:1-2.

Kim, J.H., *et al.* (2009), *Structure and dynamics of the iron-sulfur cluster assembly scaffold protein IscU and its interaction with the cochaperone HscB.* Biochemistry. 48(26):6062-6071.

Kim, J.H., *et al.* (2015), *Tangled web of interactions among proteins involved in iron-sulfur cluster assembly as unraveled by NMR, SAXS, chemical crosslinking, and functional studies.* Biochimica et Biophysica Acta. 1853(6):1416-28.

Kuzmič, P. (1996), *Program DYNAFIT for the Analysis of Enzyme Kinetic Data: Application to HIV Proteinase.* Analytical Biochemistry. 237(2):260-273.

Layer, G., *et al.* (2006), *Iron-sulfur cluster biosynthesis: characterization of Escherichia coli CYaY as an iron donor for the assembly of [2Fe-2S] clusters in the scaffold IscU.* Journal of Biological Chemistry. 281(24):16256-63.

Lesuisse, E., *et al.* (2003), *Iron use for haeme synthesis is under control of the yeast frataxin homologue (Yfh1).* Human Molecular Genetics. 12(8):879-89.

Ma, H., *et al.* (2005), *Nanoliter homogenous ultra-high throughput screening microarray for lead discoveries and IC50 profiling.* Assay and Drug Development Technologies. 3(2):177-187.

Markert, C.L. (1984), *Lactate dehydrogenase. Biochemistry and function of lactate dehydrogenase.* Cell Biochemistry and Function. 2(3):131-134.

Matos, C.A., de Macedo-Ribeiro, S., and Carvalho, A.L. (2011), *Polyglutamine diseases: the special case of ataxin-3 and Machado-Joseph disease.* Progress in Neurobiology. 95(1):26-48.

Matosevic, S., Szita, N., and Baganz, F. (2011), *Fundamentals and applications of immobilized microfluidic enzymatic reactors.* Journal of Chemical Technology & Biotechnology. 86(3):325-334.

McComb, R.B., *et al.* (1976), *Determination of the molar absorptivity of NADH.* Clinical Chemistry. 22(2):141-50.

Michaelis, L. and Menten, M. (1913), *Die Kinetik der Invertinwirkung.* Biochemische Zeitschrift. 49:333–369.

Morales, M.F. (1955), *If an Enzyme-Substrate Modifier System Exhibits Non-competitive Interaction, then, in General, its Michaelis Constant is an Equilibrium Constant.* Journal of the American Chemical Society. 77(15):4169-4170.

Morrison, J.F. (1969), *Kinetics of the reversible inhibition of enzyme-catalysed reactions by tight-binding inhibitors.* Biochimica et Biophysica Acta (BBA) - Enzymology. 185(2):269-286.

Muhlenhoff, U., *et al.* (2002), *The yeast frataxin homolog Yfh1p plays a specific role in the maturation of cellular Fe/S proteins.* Human Molecular Genetics. 11(17):2025-36.

Murie, C., *et al.* (2015), *Improving detection of rare biological events in high-throughput screens.* Journal of Biomolecular Screening. 20(2):230-241.

Nair, M., *et al.* (2004), *Solution structure of the bacterial frataxin ortholog, CyaY: mapping the iron binding sites.* Structure. 12(11):2037-48.

Nath, A. and Atkins, W.M. (2008), *A quantitative index of substrate promiscuity.* Biochemistry. 47(1):157-166.

Olp, M.D., Kalous, K.S., and Smith, B.C. (2019), *An online tool for calculating initial rates from continuous enzyme kinetic traces.* bioRxiv:700138.

Orsi, B.A. and Tipton, K.F. (1979), *Kinetic analysis of progress curves.* Methods in Enzymology. 63:159-83.

Pandolfo, M. and Pastore, A. (2009), *The pathogenesis of Friedreich ataxia and the structure and function of frataxin.* Journal of Neurology. 256 Suppl 1:9-17.

Pandya, C., *et al.* (2014), *Enzyme Promiscuity: Engine of Evolutionary Innovation.* The Journal of Biological Chemistry. 289(44):30229-30236.

Parizi, L.F., *et al.* (2018), *Peptidase inhibitors in tick physiology.* Med Vet Entomol. 32(2):129-144.

Pastore, C., *et al.* (2006), *YfhJ, a Molecular Adaptor in Iron-Sulfur Cluster Formation or a Frataxin-like Protein?* Structure. 14(5):857-867.

Pereira, C., *et al.* (2014), *Potential small-molecule activators of caspase-7 identified using yeast-based caspase-3 and -7 screening assays.* Eur J Pharm Sci. 54:8-16.

Petersen, K.J., *et al.* (2014), *Fluorescence lifetime plate reader: Resolution and precision meet high-throughput.* Review of Scientific Instruments. 85(11):113101.

Pinto, M.F. and Martins, P.M. (2016), *In search of lost time constants and of non-Michaelis-Menten parameters.* Perspectives in Science. 9:8-16.

Pinto, M.F., *et al.* (2015), *Enzyme kinetics: the whole picture reveals hidden meanings.* FEBS Journal. 282(12):2309-2316.

Pinto, M.F., *et al.* (2019), *A simple linearization method unveils hidden enzymatic assay interferences.* Biophysical Chemistry. 252:106193.

Podjarny, A., Dejaegere, A.P., and Kieffer, B., (2011), *Biophysical Approaches Determining Ligand Binding to Biomolecular Targets: Detection, Measurement and Modelling,* 1st ed. RSC Biomolecular Sciences. Royal Society of Chemistry, pp.118.

Prischi, F., *et al.* (2010), *Of the vulnerability of orphan complex proteins: The case study of the E. coli IscU and IscS proteins.* Protein Expression and Purification. 73(2):161-166.

Prischi, F., *et al.* (2010), *Structural bases for the interaction of frataxin with the central components of iron–sulphur cluster assembly.* Nature Communications. 1(1):95.

Ramazzotti, A., Vanmansart, V., and Foury, F. (2004), *Mitochondrial functional interactions between frataxin and Isu1p, the iron-sulfur cluster scaffold protein, in Saccharomyces cerevisiae.* FEBS Letters. 557(1-3):215-20.

Ramelot, T.A., *et al.* (2004), *Solution NMR Structure of the Iron–Sulfur Cluster Assembly Protein U (IscU) with Zinc Bound at the Active Site.* Journal of Molecular Biology. 344(2):567-583.

Rask-Andersen, M., Masuram, S., and Schiöth, H.B. (2014), *The Druggable Genome: Evaluation of Drug Targets in Clinical Trials Suggests Major Shifts in Molecular Class and Indication.* Annual Review of Pharmacology and Toxicology. 54(1):9-26.

Rawlings, N.D., *et al.* (2018), *The MEROPS database of proteolytic enzymes, their substrates and inhibitors in 2017 and a comparison with peptidases in the PANTHER database.* Nucleic Acids Res. 46(D1):D624-d632.

Reed, M.C., Lieb, A., and Nijhout, H.F. (2010), *The biological significance of substrate inhibition: a mechanism with diverse functions.* Bioessays. 32(5):422-9.

Rexer, T.F.T., *et al.* (2018), *One pot synthesis of GDP-mannose by a multi-enzyme cascade for enzymatic assembly of lipid-linked oligosaccharides.* Biotechnology and Bioengineering. 115(1):192-205.

Roche, B., *et al.* (2013), *Iron/sulfur proteins biogenesis in prokaryotes: formation, regulation and diversity.* Biochimica et Biophysica Acta. 1827(3):455-69.

Roskoski, R., (2014), Enzyme Assays☆. Elsevier, Chp.

Roy, A. (2018), *Early probe and drug discovery in academia: a minireview.* High-throughput. 7(1):4.

Sant'Anna, R., *et al.* (2016), *Repositioning tolcapone as a potent inhibitor of transthyretin amyloidogenesis and associated cellular toxicity.* Nature Communications. 7:10787.

Sárkány, Z., *et al.* (2019), *Chemical Kinetic Strategies for High‑Throughput Screening of Protein Aggregation Modulators.* Chemistry‑An Asian Journal. 14(4):500-508.

Scheel, H., Tomiuk, S., and Hofmann, K. (2003), *Elucidation of ataxin-3 and ataxin-7 function by integrative bioinformatics.* Human Molecular Genetics. 12(21):2845-2852.

Schnell, S. (2014), *Validity of the Michaelis-Menten equation – steady-state or reactant stationary assumption: that is the question.* The FEBS Journal. 281(2):464-472.

Schnell, S. and Hanson, S.M. (2007), *A test for measuring the effects of enzyme inactivation.* Biophys Chem. 125(2-3):269-74.

Schnell, S. and Maini, P.K. (2000), *Enzyme kinetics at high enzyme concentration.* Bulletin of mathematical biology. 62(3):483-499.

Schnell, S. and Mendoza, C. (1997), *Closed Form Solution for Time-dependent Enzyme Kinetics.* Journal of Theoretical Biology. 187(2):207-212.

Segel, L.A. (1988), *On the validity of the steady state assumption of enzyme kinetics.* Bulletin of Mathematical Biology. 50(6):579-593.

Selwyn, M.J. (1965), *A simple test for inactivation of an enzyme during assay.* Biochimica et Biophysica Acta (BBA) - Enzymology and Biological Oxidation. 105(1):193-195.

Shi, R., *et al.* (2010), *Structural basis for Fe-S cluster assembly and tRNA thiolation mediated by IscS protein-protein interactions.* PLOS Biology. 8(4):e1000354.

Shoichet, B.K. (2006), *Screening in a spirit haunted world.* Drug Discovery Today. 11(13):607-615.

Shoichet, B.K. (2006), *Screening in a spirit haunted world.* Drug Discovery Today. 11(13-14):607-615.

Siegel, L.M. (1965), *A direct microdetermination for sulphide.* Anal Biochem. 11:126-32.

Silva, A., de Almeida, A.V., and Macedo-Ribeiro, S. (2018), *Polyglutamine expansion diseases: More than simple repeats.* Journal of Structural Biology. 201(2):139-154.

Silva, A., *et al.* (2017), *Distribution of Amyloid-Like and Oligomeric Species from Protein Aggregation Kinetics.* Angewandte Chemie International Edition. 56(45):14042-14045.

Simeonov, A. and Davis, M.I., (2018), *Interference with fluorescence and absorbance.* Assay Guidance Manual [Internet]. Eli Lilly & Company and the National Center for Advancing Translational Sciences.

Sols, A. and Marco, R. (1970), *Concentrations of metabolites and binding sites. Implications in metabolic regulation.* Current Topics in Cellular Regulation. 2:227–273.

Sørensen, T.H., *et al.* (2015), *Temperature effects on kinetic parameters and substrate affinity of Cel7A cellobiohydrolases.* Journal of Biological Chemistry.

Stevens, R., Stevens, L., and Price, N.C. (1983), *The stabilities of various thiol compounds used in protein purifications.* Biochemical Education. 11(2):70-70.

Stewart, S. (2005), *A new elementary function for our curricula?* Australian Senior Mathematics Journal. 19(2).

Stroberg, W. and Schnell, S. (2016), *On the estimation errors of K_M and V from time-course experiments using the Michaelis–Menten equation.* Biophysical Chemistry. 219:17-27.

Swainston, N., *et al.* (2018), *STRENDA DB: enabling the validation and sharing of enzyme kinetics data.* FEBS Journal. 285(12):2193-2204.

Szedlacsek, S.E. and Duggleby, R.G., (1995), [6] Kinetics of slow and tight-binding inhibitors. Vol. 249. Academic Press, Chp.

Tang, Q. and Leyh, T.S. (2010), *Precise, Facile Initial Rate Measurements.* The Journal of Physical Chemistry B. 114(49):16131-16136.

Tholander, F. (2012), *Improved inhibitor screening experiments by comparative analysis of simulated enzyme progress curves.* PloS one. 7(10):e46764.

Thompson, R.E., *et al.* (2017), *Tyrosine sulfation modulates activity of tick-derived thrombin inhibitors.* Nature Chemistry. 9:909.

Thorne, N., Auld, D.S., and Inglese, J. (2010), *Apparent activity in high-throughput screening: origins of compound-dependent assay interference.* Current Opinion in Chemical Biology. 14(3):315-324.

Tipton, K.F., *et al.* (2014), *Standards for Reporting Enzyme Data: The STRENDA Consortium: What it aims to do and why it should be helpful.* Perspectives in Science. 1:131–137.

Tirat, A., *et al.* (2005), *Synthesis and characterization of fluorescent ubiquitin derivatives as highly sensitive substrates for the deubiquitinating enzymes UCH-L3 and USP-2.* Analytical Biochemistry. 343(2):244-255.

Turrens, J.F. and Boveris, A. (1980), *Generation of superoxide anion by the NADH dehydrogenase of bovine heart mitochondria.* Biochemical Journal. 191(2):421.

Tzafriri, A.R. (2003), *Michaelis-Menten kinetics at high enzyme concentrations.* Bulletin of mathematical biology. 65(6):1111-1129.

van Doorn, J., *et al.* (2018), *Bayesian Inference for Kendall's Rank Correlation Coefficient.* The American Statistician. 72(4):303-308.

Walker, A.C. and Schmidt, C.L.A. (1944), *Studies on histidase.* Archives of Biochemistry and Biophysics. 5:445-467.

Ware, F.L. and Luck, M.R. (2017), *Evolution of salivary secretions in haematophagous animals.* Bioscience Horizons. 10:hzw015.

Watson, E.E., *et al.* (2019), *Rapid assembly and profiling of an anticoagulant sulfoprotein library.* Proceedings of the National Academy of Sciences. 116(28):13873.

Wu, G., Yuan, Y., and Hodge, C.N. (2003), *Determining appropriate substrate conversion for enzymatic assays in high-throughput screening.* Journal of Biomolecular Screening. 8(6):694-700.

Xie, X.S. (2013), *Enzyme Kinetics, Past and Present.* Science. 342(6165):1457.

Yan, R., *et al.* (2013), *Ferredoxin competes with bacterial frataxin in binding to the desulfurase IscS.* Journal of Biological Chemistry. 288(34):24777-87.

Yang, J., Copeland, R.A., and Lai, Z. (2009), *Defining balanced conditions for inhibitor screening assays that target bisubstrate enzymes.* Journal of Biomolecular Screening. 14(2):111-20.

Yoon, T. and Cowan, J.A. (2003), *Iron-sulfur cluster biosynthesis. Characterization of frataxin as an iron donor for assembly of [2Fe-2S] clusters in ISU-type proteins.* Journal of the American Chemical Society. 125(20):6078-84.

Yoon, T. and Cowan, J.A. (2004), *Frataxin-mediated iron delivery to ferrochelatase in the final step of heme biosynthesis.* Journal of Biological Chemistry. 279(25):25943-6.

Yu, K., *et al.* (2011), *A high-throughput colorimetric assay to measure the activity of glutamate decarboxylase.* Enzyme and Microbial Technology. 49(3):272-276.

Zhang, J.-H., Chung, T.D., and Oldenburg, K.R. (1999), *A simple statistical parameter for use in evaluation and validation of high throughput screening assays.* Journal of biomolecular screening. 4(2):67-73.

Zhong, X. and Pittman, R.N. (2006), *Ataxin-3 binds VCP/p97 and regulates retrotranslocation of ERAD substrates.* Human Molecular Genetics. 15(16):2409-2420.

Zhu, A., Romero, R., and Petty, H.R. (2010), *A sensitive fluorimetric assay for pyruvate.* Analytical Biochemistry. 396(1):146-51.

Zimmerle, C.T. and Frieden, C. (1989), *Analysis of progress curves by simulations generated by numerical integration.* The Biochemical journal. 258(2):381-387.

Annexes

Annex 1 – <u>Article I (Original research article published)</u>
Pinto, M.F. and Martins, P.M. (2016), In search of lost time constants and of non-Michaelis-Menten parameters. Perspectives in Science. 9:8-16. doi:10.1016/j.pisc.2016.03.024

Annex 2 – <u>Article II (Original research article published)</u>
Pinto, M.F., Ripoll-Rozada, J., Ramos, H., Watson, E.E., Franck, C., Payne, R.J., Saraiva, L., Pereira, P.J.B., Pastore, A., Rocha, F., Martins, P.M. (2019), A simple linearization method unveils hidden enzymatic assay interferences. Biophysical Chemistry. 252:106193. doi:10.1016/j.bpc.2019.106193

Available online at www.sciencedirect.com

ScienceDirect

journal homepage: www.elsevier.com/pisc

In search of lost time constants and of non-Michaelis—Menten parameters[☆]

Maria F. Pinto[a,b], **Pedro M. Martins**[a,b,*]

[a] *ICBAS, Instituto de Ciências Biomédicas Abel Salazar da Universidade do Porto, Rua de Jorge Viterbo Ferreira n°. 228, 4050-313 Porto, Portugal*
[b] *LEPABE, Laboratório de Engenharia de Processos, Ambiente, Biotecnologia e Energia, Departamento de Engenharia Química, Faculdade de Engenharia da Universidade do Porto, Rua Dr. Roberto Frias, 4200-465 Porto, Portugal*

Received 5 February 2016; accepted 1 March 2016
Available online 27 August 2016

KEYWORDS

Enzyme kinetics;
Michaelis—Menten equation;
Mathematical model;
Renewal theory;
Non-Michaelis—Menten enzyme kinetics

Summary   Upon completing 100 years since it was published, the work *Die Kinetik der Invertinwirkung* by Michaelis and Menten (MM) was celebrated during the 6th Beilstein ESCEC Symposium 2013. As the 7th Beilstein ESCEC Symposium 2015 debates enzymology in the context of complex biological systems, a post-MM approach is required to address cell-like conditions that are well beyond the steady-state limitations. The present contribution specifically addresses two hitherto ambiguous constants whose interest was, however, intuited in the original MM paper: (i) the characteristic time constant τ_∞, which can be determined using the late stages of any progress curve independently of the substrate concentration adopted; and (ii) the dissociation constant K_S, which is indicative of the enzyme—substrate affinity and completes the kinetic portrayal of the Briggs—Haldane reaction scheme. The rationale behind τ_∞ and K_S prompted us to revise widespread concepts of enzyme's efficiency, defined by the specificity constant k_{cat}/K_M, and of the Michaelis constant K_M seen as the substrate concentration yielding half-maximal rates. The alternative definitions here presented should help recovering the wealth of published k_{cat}/K_M and K_M data from the criticism that they are subjected. Finally, a practical method is envisaged for objectively determining enzyme's activity, efficiency and affinity — $(EA)^2$ — from single progress curves. The $(EA)^2$ assay can be conveniently applied even when the concentrations of substrate and enzyme are not accurately known.

* Corresponding author at: ICBAS, Instituto de Ciências Biomédicas Abel Salazar da Universidade do Porto, Rua de Jorge Viterbo Ferreira n°. 228, 4050-313 Porto, Portugal.
E-mail addresses: maria.filipa.s.pinto@gmail.com (M.F. Pinto), pmartins2106@gmail.com (P.M. Martins).

http://dx.doi.org/10.1016/j.pisc.2016.03.024

Introduction

The year of 2013 marked the one hundredth anniversary of the publication of the classic Michaelis and Menten (MM) paper *Die Kinetik der Invertinwirkung* (Michaelis and Menten, 1913), which became the standard approach to quasi-steady-state (QSS) enzyme kinetics. Supported by the work of earlier authors, most notably Brown (1902) and Henri (1902, 1903), MM understood the significance of pH control in enzymatic experiments and acknowledged that initial rates were easier to interpret than time courses as they are not restrained by issues such as the reverse reaction, product inhibition or enzyme inactivation (Cornish-Bowden, 2012). Modern representations of the MM model use the Briggs and Haldane reaction scheme encompassing the reversible combination of free enzyme E and substrate S to form the enzyme–substrate complex ES followed by its irreversible transformation into product P and release of enzyme (Eq. (1)) (Briggs and Haldane, 1925)

$$E + S \underset{k_{-1}}{\overset{k_1}{\rightleftharpoons}} ES \overset{k_2}{\longrightarrow} E + P \tag{1}$$

where k_1 and k_{-1} are the rate constants of the reversible binding step and k_2 is the rate constant of the catalytic step. The evolution of the concentration of the different species with time t is mathematically described by the following system of first-order differential equations

$$\frac{d[S]}{dt} = -k_1[E][S] + k_{-1}[ES] \tag{2}$$

$$\frac{d[ES]}{dt} = k_1[E][S] - (k_{-1} + k_2)[ES] \tag{3}$$

$$\frac{d[E]}{dt} = -k_1[E][S] + k_{-1}[ES] + k_2[E] \tag{4}$$

$$\frac{d[P]}{dt} = k_2[ES] \tag{5}$$

subject to the initial conditions ($[S]$, $[E]$, $[ES]$, $[P]$) = $(S_0, E_0, 0, 0)$. Although the analytical solution of Eqs. (2)–(5) is not known (Berberan-Santos, 2010), a simplified alternative results from adopting the QSS approximation stating that, in the presence of a large excess of substrate, the concentration of the enzyme–substrate complex remains constant after the initial ES build-up period has ended (Briggs and Haldane, 1925). If, in addition, the duration of the transient period is short enough to assume invariant $[S]$, the reactant stationary approximation is applicable (Hanson and Schnell, 2008), and the final form of the MM equation is obtained (Eq. (6))

$$v_0 = \frac{V_{max}S_0}{K_M + S_0} \tag{6}$$

with v_0 being the initial reaction rate; V_{max}, the limit reaction rate obtained for very high substrate concentration values; and K_M, the Michaelis constant. In the Briggs and Haldane notation V_{max} corresponds to k_2E_0 and K_M corresponds to $(k_{-1} + k_2)/k_1$; in practical terms, V_{max} is written as $k_{cat}E_0$ to extend its use to reaction schemes of higher complexity than Briggs and Haldane's, while K_M is commonly referred as the concentration of substrate for which $v_0 = 0.5V_{max}$. The QSS and the reactant stationary approximations severely limit

the applicability of the MM equation to the initial phases of enzymatic reactions that start with great substrate excess over the enzyme ($S_0 \gg E_0$) (Pinto et al., 2015; Segel, 1988; Hanson and Schnell, 2008). With the publication of the Pinto et al. (PEA) model in 2015, additional threats associated to the usage of the classical formalism were identified, at the same time that the "whole picture" of single active-site enzyme kinetics without inhibition was revealed (Pinto et al., 2015). The PEA model also uncovered new applications or "hidden meanings" in the Briggs and Haldane mechanism, of which the present contribution particularly focuses the cases of the characteristic time constant τ_∞ and of the dissociation constant K_S. These parameters were chosen as they help to answer some of the new problems posed by Systems Biology while studying increasingly realistic enzymatic networks. Not only that the following sections illustrate how τ_∞ and K_S can be used to characterize enzymatic activity, enzymatic efficiency and enzyme–substrate affinity in a straightforward and unambiguous manner.

Numerical procedures

The system of differential equations describing the Briggs and Haldane reaction scheme (Eqs. (2)–(5)) was expressed in normalized units as Eqs. (7)–(9) (Pinto et al., 2015)

$$-\left(1 - \frac{K_S}{K_M}\right) \frac{ds}{d\theta} = e_0 s - c\left(\frac{K_S}{K_M} + s\right) \tag{7}$$

$$\left(1 - \frac{K_S}{K_M}\right) \frac{dc}{d\theta} = e_0 s - c(1 + s) \tag{8}$$

$$\frac{dp}{d\theta} = c \tag{9}$$

where $s = [S]/K_M$, $c = [ES]/K_M$, $p = [P]/K_M$, $\theta = k_2 t$ and $K_S = k_{-1}/k_1$. Enzymatic reaction progress curves showing the evolution of scaled product concentration p over scaled time θ were simulated with Mathworks® MATLAB R2013b. A script was developed to this end in which a MATLAB ordinary differential equation (ODE) solver was employed to numerically solve Eqs. (7)–(9) over the scaled time. The specific ODE solver used to this effect was ode45, a one-step solver (i.e. when computing the solution for t_n, the solver only requires the solution at the immediately preceding time point, t_{n-1}) based on an explicit Runge–Kutta(4,5) formula, the Dormand-Prince pair (Dormand and Prince, 1980). Numerical solutions were obtained over different ranges of integration of θ for limiting values of the scaled dissociation constant K_S/K_M and for different sets of e_0 and s_0 initial conditions.

Results

The characteristic time constant (τ_∞) and the enzyme efficiency

The analytical solution describing single active-site enzyme kinetics without inhibition was obtained after introducing the "pivotal variable" $(S_0 - P)/v$ representing the concentration of product still to be formed $(S_0 - P)$ over the instant reaction rate v (Pinto et al., 2015). Fig. 1A illustrates the

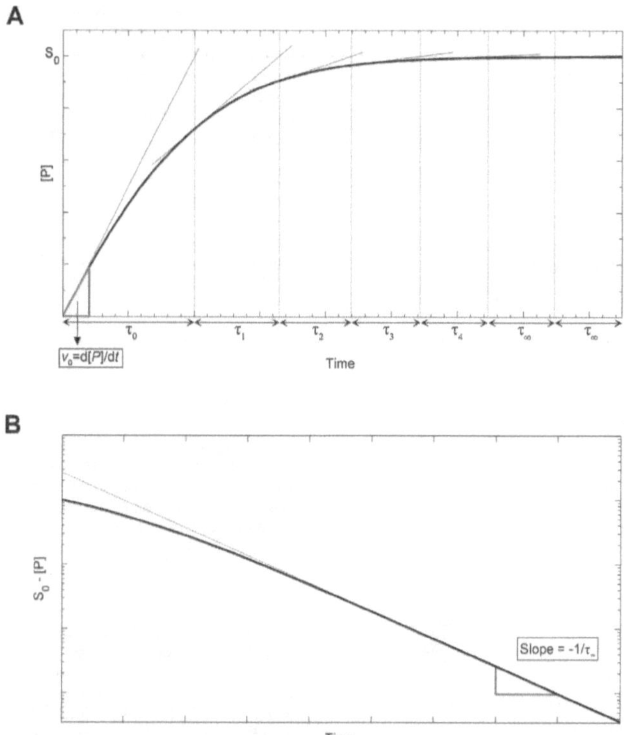

Figure 1   Different representations of the theoretical progress curve obtained from the numerical solution of the ODE system comprising Eqs. (7)–(9) using $S_0/K_M = 1$, $E_0/K_M = 0.01$ and $K_S/K_M = 1$. (A) Product concentration [P] represented over time t in a linear plot. Red tangent lines represent the period of time τ_n that would be required to complete the reaction if the instant reaction rates were maintained. For long reaction times this period of time tends to the value of the characteristic time constant τ_∞. The slope of the initial tangent corresponds to the value of the initial reaction rate v_0. (B) Log-linear plot of the concentration of product still to be formed ($S_0 - P$) as a function of time. The slope of final tangent (red dashed line) corresponds to the negative reciprocal of the characteristic time constant.

physical meaning of the pivotal variable as the period of time τ_n that would be required to complete the reaction if the instant reaction rate was maintained. Alternatively, the negative reciprocal of this variable is promptly computed as the instantaneous slope of the ($S_0 - P$) time–course curve represented in a log-linear scale (Fig. 1B). The asymptotic limit of ($S_0 - P$)/v for late reaction phases is here defined as the characteristic time constant τ_∞ and corresponds to the reciprocal of the "integration constant" shown in the original MM paper to be independent of the initial substrate concentration (Michaelis and Menten, 1913). Later interpretation of QSS results identified the integration constant as the specificity constant k_2/K_M (or, more generically, k_{cat}/K_M) multiplied by the enzyme concentration (Johnson and Goody, 2011), while its reciprocal corresponds to the period of time τ needed to completely exhaust the existing

substrate if the initial reaction rate is maintained and the enzyme is operating under first-order conditions (Cornish-Bowden, 1987). Despite the similarities between the latter definition and our own definition of τ_∞, the following differences should be noted: the time constant τ is defined in relation to the initial reaction rates under QSS conditions, whereas τ_∞ is concerned with the late reaction phases under whatever experimental conditions. From the definition of the pivotal variable for long reaction times given in the Supporting Information of the PEA paper (Pinto et al., 2015), the following relationship exists between τ_∞ and τ (Eq. (10)):

$$\tau_\infty = \frac{\tau}{2}\left(1 + e_0 + \sqrt{(1 + e_0)^2 - 4\left(1 - \frac{K_S}{K_M}\right)e_0}\right) \qquad (10)$$

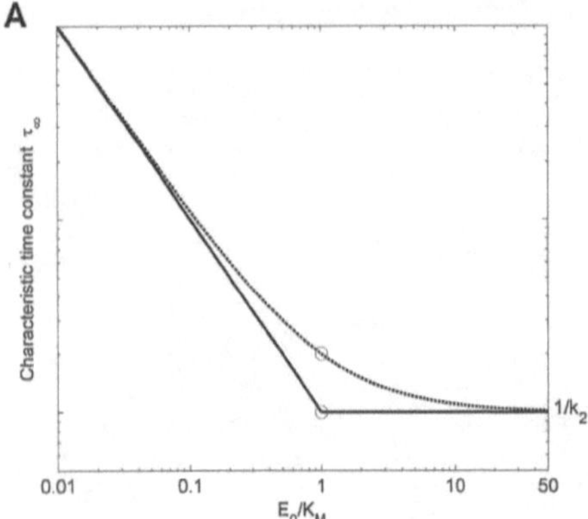

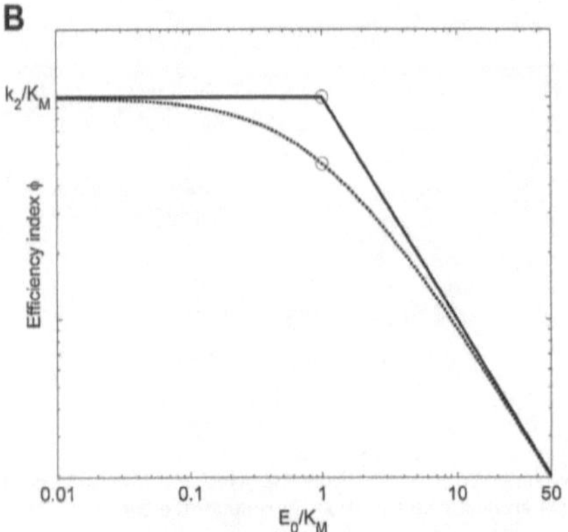

Figure 2 The characteristic time constant and enzyme efficiency. Log–log plots depicting the influence of the K_M-normalized enzyme concentration on the (A) characteristic time constant τ_∞ and on the (B) efficiency index ϕ for limiting values of the scaled dissociation constant $K_S/K_M = 0$ (solid lines) and $K_S/K_M = 1$ (dashed lines). The blue round markers show the point where the largest difference between both curves is observed. (A) The smallest value for the characteristic time is obtained for $E_0 > K_M$. (B) The maximal efficiency index ϕ_{max} is obtained for $E_0 < K_M$.

The representation of this function in Fig. 2A takes into account the alternative definition of $1/\tau$ as $k_2 e_0$ to show that the shortest characteristic time corresponds to $1/k_2$ and is obtained for enzyme concentrations above the Michaelis constant. This compromise between finishing reaction rates and enzyme concentration motivated us to propose an efficiency index ϕ balancing kinetic performance over the enzyme expenditure:

$$\phi = \frac{1/\tau_\infty}{E_0} \tag{11}$$

Defined in this way, enzyme efficiency is exempted from the practical limitations of the specificity constant, whose application to compare the catalytic efficiency of different enzymes in the catalysis of the same substrate has been discouraged (Eisenthal et al., 2007). In fact, by attending to the final phases of the enzymatic reaction, the definition of ϕ is free from the ambiguities caused by the role of the substrate concentration on the initial reaction rates (Eisenthal et al., 2007). On the other hand, the fact illustrated in Fig. 2B that the maximum value of efficiency ϕ_{max} corresponds to the value of k_2/K_M (or, more generically to k_{cat}/K_M), might be extremely convenient so as to recover published k_{cat}/K_M data from any misgivings while comparing the efficiency of different enzymes. Finally, and as addressed more in detail in the discussion section, the efficiency index can be straightforwardly estimated from a single enzymatic assay using Eq. (11) and the values of τ_∞ determined as described in Fig. 1B.

The K_S/K_M ratio and the enzyme–substrate affinity

In the original MM paper, the now-called Michaelis constant K_M was defined as the protein–ligand dissociation constant (Michaelis and Menten, 1913), which for enzyme–substrate complexes is now commonly represented by K_S. Comparing their mathematical formulations given in the introduction part shows that the catalytic step (rate constant k_2) must be much slower than the unbinding step (rate constant k_{-1}) for K_M to be equivalent to K_S (Baici, 2015). In the PEA paper, K_S is referred to as a non-MM constant, which, together with K_M and V_{max}, completes the portrayal of the 3-parameter mechanism proposed by Briggs and Haldane (Pinto et al., 2015). Fig. 3 shows two sets of theoretical curves simulated for enzyme concentrations much lower than K_M (Fig. 3A) and equal to K_M (Fig. 3B) to illustrate the peculiar role of K_S in both situations. Fig. 3A partly explains the absence of K_S from QSS kinetic analysis, seeing that the enzyme–substrate affinity has a weak effect on the progress curves, which is only visible for product conversions below 5%, and considering substrate concentrations S_0 close to E_0. This does not mean that K_S is equivalent to K_M, only that the effect of K_S is masked under conditions of great substrate excess. In the other extreme, experimental conditions for which the enzyme concentration is of the same order of magnitude of K_M (and $S_0 \leq E_0$) are expected to clearly reveal the effect of K_S during initial and late phases of the progress curves (Pinto et al., 2015); for this reason, and because of the biological interest, this is considered a "critical region of conditions" that is potentially representative of an intracellular environment (Schnell and Maini, 2000; Tzafriri, 2003; Bersani and Dell'Acqua, 2011). Fig. 3B shows that asymptotically high affinities between enzyme and substrate ($K_S/K_M = 0$) should produce characteristic product accumulation curves with sigmoidal (rather than hyperbolic/linear) onsets. Since low K_S/K_M ratios mean much faster product formation rates than enzyme–substrate dissociation rates, it might be technically difficult to access the earlier phases of such kinetic curves and discern their shape, especially when high enzyme concentrations are involved. The PEA alternative to estimate the value of K_S/K_M is through the characteristic time constant τ_∞, which, as described in the previous subsection, can be straightforwardly obtained from a single enzymatic assay.

                                                                    M.F. Pinto, P.M. Martins

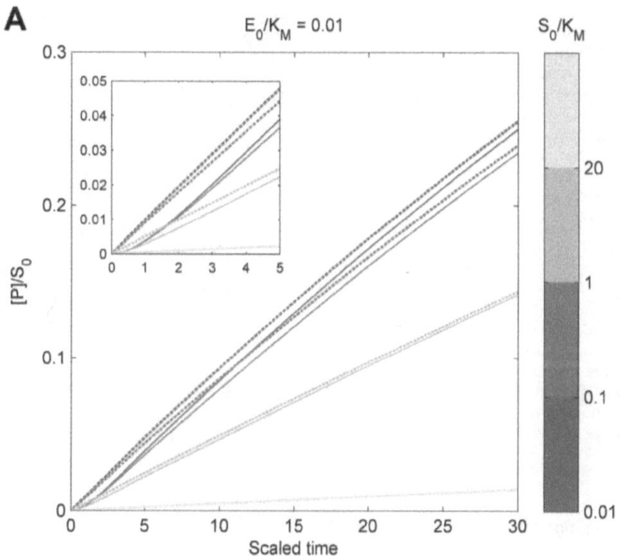

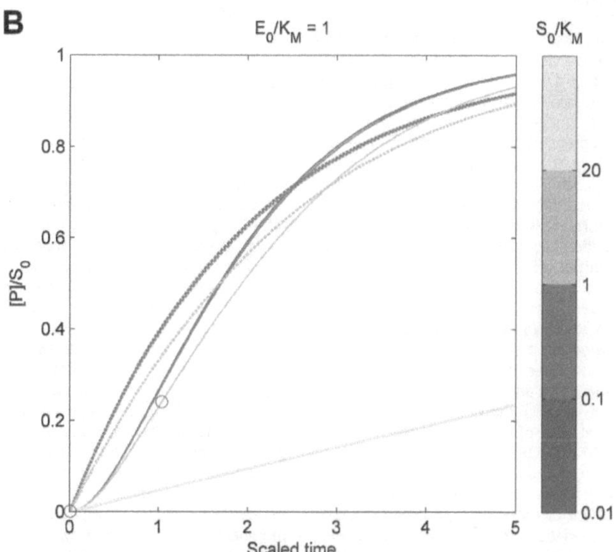

Figure 3 Major differences between theoretical progress curves calculated for limiting values of the dissociation constant $K_S/K_M = 0$ (solid lines) and $K_S/K_M = 1$ (dashed lines). Progress curves represented as the linear plots of the normalized product concentration $[P]/S_0$ over the scaled time $\theta = k_2 t$. The system of ODE comprising Equations 7–9 was solved using the set of S_0/K_M values indicated in the log-scaled color bars for (A) $E_0/K_M = 0.01$ and (B) $E_0/K_M = 1$. (B) The blue round markers on the curves obtained

Given that the characteristic time constant is independent of the initial substrate concentration, values of S_0 as high as the solubility limit can be adopted in order to extend the duration of the catalytic reactions over technically accessible time periods. As previously represented in Fig. 2A, the influence of E_0 on τ_∞ is not significantly affected by the value of the K_S/K_M ratio, unless enzyme concentrations close to K_M are considered. This window of conditions is, therefore, recommended to estimate the dissociation constant from experimentally determined characteristic time constants. The K_S/K_M value follows directly from Eq. (10) rewritten as Eq. (12)

$$\frac{K_S}{K_M} = 1 + (k_{cat}\tau_\infty)^2 \frac{E_0}{K_M} - (k_{cat}\tau_\infty)\left(1 + \frac{E_0}{K_M}\right) \qquad (12)$$

which requires previous estimations of the MM parameters using, for example, the PEA model equations (14) in the appendix section or the MM equation (Eq. (6) for QSS conditions only). In the Discussion section we anticipate some of the practical and fundamental consequences arising from the accurate knowledge of the parameter K_S.

Discussion

The present work is the first follow-up of the PEA model, which, as the acronym incidentally suggests, is envisaged to seed several other future applications in modern enzymology. Specifically, we took the opportunity at the 7th Beilstein ESCEC Symposium to expand the meaning and practical significance of the characteristic time constant τ_∞ and of the equilibrium dissociation constant K_S. The relevance of these parameters was already intuited in the 1915 paper of MM, seeing that $1/\tau_\infty$ and K_S correspond, in the limit cases, to the original "integration constant" and to the Michaelis constant, respectively. More than enlarging the QSS scope, our approach motivates a renewed interpretation of the fundamental meaning of MM and non-MM kinetic constants. For example, enzyme efficiency defined in relation to τ_∞ is not affected by the concentration of substrate and, therefore, it is free from the ambiguities associated to the specificity constant defined as the k_{cat}/K_M ratio extracted from initial velocity experiments (Eisenthal et al., 2007). A direct indicator of the enzyme's kinetic performance, the value of $1/\tau_\infty$ is also an apparent first-order rate constant that increases with the concentration of enzyme until the upper limit of k_2 is attained for $E_0 > K_M$ (Fig. 2A). Consequently, enzyme efficiency is here presented as the kinetic performance balanced over the total enzyme expenditure $\phi = 1/(\tau_\infty E_0)$. Fig. 2B showed that the efficiency index reaches a maximal value of k_2/K_M that is nearly invariant for enzyme concentrations below K_M. This value of ϕ_{max}, which operationally corresponds to k_{cat}/K_M, can be used to compare the catalytic effectiveness of different enzymes for technological applications or for enzyme evolution studies. As the differences summarized in Table 1 intend to illustrate, the numerical equivalence between ϕ_{max} and the specificity constant is circumstantial and does not imply a common underlying principle. Different fundamental definitions (#4 and #5 in Table 1) stipulate different methodological procedures for the determination of the two indicators (#1 to #3 in Table 1) which, nevertheless, should produce the same

Table 1 Different interpretations of k_{cat}/K_M in the light of the MM model (as a specificity constant) and in the light of the PEA model (as the maximal enzyme efficiency ϕ_{max}). Differences 1–3 concern parameter estimation methodologies; differences 4 and 5 concern kinetic and operational meanings, respectively; difference 6 concerns reaction schemes other than Briggs and Haldane's.

#	Specificity constant	ϕ_{max}
1	Estimated based on *initial* reaction rates v_0	Estimated based on the characteristic time constant τ_∞ during *late* reaction phases
2	*Limited* to QSS experimental conditions	Estimations of ϕ are *not limited* to any experimental condition; ϕ_{max} is reached for $E_0 < K_M$
3	Substrate concentration *influences* the v_0-based enzyme's efficiency (Eisenthal et al., 2007)	Substrate concentration *does not* influence τ_∞-based enzyme's efficiency
4	Corresponds to an apparent second-order rate constant	Corresponds to an apparent first-order rate constant expressed per units of enzyme concentration
5	Sets the *lower* limit for enzyme-substrate association rate constant (Fersht, 1999)	Sets the *upper* limit of the ratio enzyme performance/enzyme expenditure
6	It is *not* affected by product inhibition	It may be affected by product inhibition

numerical results, provided that the Briggs and Haldane mechanism holds true. Reaction schemes involving product inhibition may originate different values of k_{cat}/K_M if estimated as ϕ_{max} or as a specificity constant (#6 in Table 1). Although product inhibition is not contemplated by the PEA model, the common usage of apparent rate constants (such as k_{cat}) as an approximation to true rate constants (such as k_2) might also be extended to the efficiency index, whose apparent value may help to characterize quantitatively the deviations from Briggs and Haldane kinetics.

Another MM parameter subject to a renewed PEA perspective is the Michaelis constant itself. Appointed as less important than the parameters k_{cat} and k_{cat}/K_M (Johnson and Goody, 2011), the value of K_M is frequently defined as the concentration of substrate producing $v_0 = 0.5V_{max}$; on the other hand, the formulation $K_M = (k_{-1} + k_2)/k_1$ indicates that the Michaelis constant is an overall/apparent dissociation constant of all enzyme-bound species (Fersht, 1999). The latter definition is directly concerned with the enzyme—substrate affinity, which can be characterized accurately using true dissociation constants (K_S) determined as described in the previous subsection. The PEA model additionally shows that the first definition of K_M (as the substrate concentration yielding half-maximal rates) loses its validity outside the region of QSS conditions (Pinto et al., 2015). For example, for $E_0 > S_0$ the initial reaction rate v_0

becomes linear dependent on the substrate concentration in the cases of very low enzyme–substrate affinity ($K_S/K_M \sim 1$) – see Eq. (14b) in the appendix section. Instead, Fig. 2 confers to parameter K_M the biophysical significance of a threshold enzyme concentration. According to Fig. 2A, K_M is the smallest enzyme concentration required to achieve the shortest completion time, i.e. required to conclude the enzymatic reaction at the fastest rates. Perhaps more useful for *in vivo* and *in vitro* kinetic analysis, Fig. 2B presents K_M as the maximum enzyme concentration that can be kept without losing catalytic efficiency – after this limit, increasing enzyme expenditure no longer accelerates the concluding reaction phases. Curiously enough, enzyme concentrations close to the value of K_M are also the most favorable to experimentally investigate the effect of the enzyme–substrate dissociation constant on the characteristic time constant (Eq. (10)). According to this new angle of approach, enzyme efficiency can be regulated by dynamically controlling the enzyme's abundance in the cell. Concentration levels close to the reference value of K_M are important for the enzyme to be critically sensible to the structural affinity of different metabolites. By systematically adopting QSS conditions, it is conceivable that *in vitro* enzymatic assays have been missing kinetic aspects of metabolic homeostasis that are important (Pinto et al., 2015), for example, in molecular systems biology (Finn and Kemp, 2014) and in drug discovery (Acker and Auld, 2014; Yang et al., 2009; Sols and Marco, 1970).

The enzyme–substrate affinity is important to define which catalysis occurs preferentially in a cellular environment crowded with multiple enzymes and substrates that possibly act as competitors toward each other. Therefore, the explanation for the apparent disregards of the dissociation constant K_S compared to k_{cat} or K_M resides in the lack of straightforward methods to estimate this non-MM constant. Existing methods for the determination of all individual rate constants require specific techniques designed to measure transient-state kinetics, the interpretation of which is not exempted from simplifying hypothesis such as the reactant stationary approximation during the pre-steady-state phases (Hanson and Schnell, 2008; Fersht, 1999) or the linearization of the reaction mechanism for time–relaxation analysis (Cornish-Bowden, 2012). These limitations are not present in the PEA method for the determination of K_S using the characteristic time constant and Eq. (12). By facilitating the characterization of enzyme specificity, we also expect to contribute to the understanding of enzyme evolution and enzyme promiscuity, upon which the design of novel biological functions is based (Pandya et al., 2014). A quantitative description of the enzyme response to alternative substrates is now possible using true dissociation constants as an alternative to entropic predictions based on the k_{cat}/K_M ratio (Nath and Atkins, 2008).

A single assay to estimate enzyme activity, efficiency and affinity (EA)^2

Estimating the MM parameters requires different enzymatic reactions to be carried out adopting substrate concentrations S_0 above and below K_M and in great excess over the enzyme ($S_0 \gg E_0$). Although the usage of a single progress curve to determine K_M and V_{max} is theoretically possible,

this procedure is discouraged in practice in view of the undefined time span over which the QSS approximation is valid (Duggleby, 2001). The insights provided by the PEA model let us envisage a new method to determine the classic parameters from a single enzymatic reaction and in an unbiased manner. In addition, the information thus, obtained can be used to analyze a second progress curve to estimate the non-MM parameter K_S. Because this method characterizes enzyme activity, efficiency and affinity we call it the (EA)^2 assay. In principle, the (EA)^2 assay involves the following steps:

1. Measure the progress curve of the enzymatic reaction under typical QSS conditions ($S_0 \gg E_0$).
2. Determine the initial reaction rate v_0 as indicated in Fig. 1A.
3. Determine the characteristic time constant τ_∞ as indicated in Fig. 1B. Assume that $\tau_\infty = \tau$.
4. Estimate V_{max} from Eq. (6) (Equation MM) rewritten as $V_{max} = v_0 S_0/(S_0 - v_0\tau)$.
5. Estimate $K_M = V_{max}\tau$.
6. The condition $\tau_\infty = \tau$ in step 3 is only valid for $E_0 \ll K_M$ (Fig. 2A). Check if $E_0 < 0.1K_M$.
6.1. If not, restart with a more diluted enzyme solution.
7. Estimate enzyme's activity as V_{max}/E_0 (equivalent to k_{cat}).
8. Estimate the maximal enzyme's efficiency $\phi_{max} = 1/(\tau_\infty E_0)$ (corresponding to k_{cat}/K_M for the conditions of step 6).
9. Measure a new progress curve adopting $E_0 = K_M$ and determine a new value of τ_∞.
10. Estimate the dissociation constant K_S characterizing the enzyme–substrate affinity. Use the value of τ_∞ estimated in step 9 and Eq. (12) rewritten as $K_S/K_M = (1 - k_{cat}\tau_\infty)^2$.

Notably, this method does not require to know an accurate value of the substrate concentration S_0, provided that this value is assuredly much higher than the product $v_0\tau_\infty$ so as to obtain $V_{max} = v_0$ in step 4. The MM parameters can alternatively be determined using the PEA model in Eqs. (14) in the appendix section or the MM equation (Eq. (6) for QSS conditions only). When the enzyme molarity is not accurately known, the (EA)^2 assay might also be useful to estimate the lower limit of the catalytic power taking into consideration that E_0 estimates such as absorbance readings at 280 nm are in excess, thus, yielding lower limits of enzyme activity V_{max}/E_0 and of enzyme efficiency $1/(\tau_\infty E_0)$. In another instance, if only the amount of impure powdered enzyme is known, enzyme efficiency can be expressed in units of s^{-1}(mg/l)$^{-1}$ as an alternative to s^{-1}M^{-1}, similarly to what happens with the catalytic activity expressed as the amount of enzyme converting the substrate into product at a given rate (1 mol/s or 1 μmol/min for katal or international unit IU, respectively). It may occur that the (EA)^2 assay fails to produce useful data because of either too slow or too fast enzymatic reactions; in the first case, sample conditions may not be maintained with time (e.g. protein degradation leading to enzyme-activity loss); in the latter case, the reaction may finish before any valid measurement is performed – especially under the $E_0 = K_M$ conditions of step 9. The solution to these problems involves decreasing or increasing of the substrate concentration values within the operational limits to prolong or shorten the reaction span to convenient limits. Obtaining enzyme samples as concentrated as the K_M-order of magnitude might also not be possible in practice. In those cases, the estimation of the K_S/K_M ratio is still possible using the initial phases of the progress curves measured using dilute enzyme solutions (Pinto et al., 2015). During the application of the PEA model and, in particular, of the $(EA)^2$ assay, the Briggs and Haldane mechanism is implicitly assumed to be valid. As previously discussed in Table 1, deviations from this mechanism can be identified by comparing the estimations of MM parameters obtained from initial and late phases of the enzymatic reactions using, in one case, Equation 14 in the appendix section, and in the other, the characteristic time constant τ_∞. We intend to keep developing the ideas organized in this paper by applying them on the characterization of enzymatic systems with biological and industrial interest.

Conclusion

Firstly published in the same year of the classic MM paper, Marcel Proust's novel *À la Recherche du Temps Perdu*, In Search of Lost Time (1913–1927), gives the motif for the title of the present contribution, in which we try to recuperate the fundamental meanings of the characteristic time constant τ_∞ and of the equilibrium dissociation constant K_S. This exercise is based on the recently published PEA model that provides, after a long wait, the closed-form solution of the Briggs and Haldane kinetic mechanism (Pinto et al., 2015). Although the Briggs and Haldane mechanism is the minimal reaction scheme needed to explain enzyme catalysis, it remained very incompletely described by the existing analytical solutions. The pivotal variable of the PEA model measured for late reaction phases gives a practical estimate of the characteristic time constant τ_∞, which in turn is helpful to clarify the concepts of enzyme efficiency and selectivity. The maximal enzyme efficiency ϕ_{max} corresponds to the value of $1/(\tau_\infty E_0)$ measured for concentrations of enzyme below K_M (Fig. 2B). Parameter ϕ_{max} is expected to help in recovering the wealth of published k_{cat}/K_M data from the criticism it has been voted as an efficiency standard: although both parameters are, in most cases, numerically equivalent, ϕ_{max} is free from the conceptual limitations of k_{cat}/K_M (Table 1). The PEA framework also provides a renewed perspective of the somewhat obscure Michaelis constant K_M as a threshold enzyme concentration above which the catalytic efficiency starts to decrease. The practical definition of K_M as the substrate concentration yielding half-maximal rates should be adopted carefully as it loses accuracy under non-QSS conditions. The true dissociation constant K_S can now be straightforwardly determined from a single progress curve without requiring specific experimental arrangements or model simplifications. Besides completing the Briggs and Haldane portrayal of the catalytic cycle, this parameter objectively characterizes the affinity of the enzyme to different substrates, thus, contributing to the study of enzyme evolution and promiscuity. Summarizing our conclusions, a practical method to determine enzyme activity, efficiency and affinity from single progress curves is proposed, in which model parameters

are rapidly estimated even if the concentrations of substrate and enzyme are not accurately known.

Acknowledgements

We thank Antonio Baici for his invaluable comments and suggestions and appreciate his interest in our manuscript. This work was financially supported by: Project UID/EQU/00511/2013-LEPABE (Laboratory for Process Engineering, Environment, Biotechnology and Energy–EQU/00511) by FEDER funds through Programa Operacional Competitividade e Internacionalização — COMPETE2020 and by national funds through FCT - Fundação para a Ciência e a Tecnologia; MFP gratefully acknowledges grant no. SFRH/BD/109324/2015 from FCT, Portugal (Programa Operacional Capital Humano (POCH), UE).

Appendix A. Appendix

The following equations comprise the overall and stationary formulations of the PEA model as described by Pinto et al. in 2015. The overall analytical solution corresponds to Eq. (13a), where scaled variables are used, namely $\tau = K_M/V_{max}$, $e_0 = E_0/K_M$, $s_0 = S_0/K_M$, $\theta = t/(e_0\tau)$ and $\beta = 1 - K_S/K_M$.

$$\frac{S_0 - P}{v} = \frac{\tau}{2}\left(1 + e_0 + \tilde{s} + \frac{\tilde{\lambda}}{\tanh(\tilde{\lambda}\theta/2\beta)}\right) \tag{13a}$$

The corresponding daughter variables $\tilde{s}$, $\tilde{\lambda}$, s^* and θ^* are given by Eqs. (13b)–(13e). The value of λ^* in Eq. (13e) corresponds to the value of $\tilde{\lambda}$ calculated by Eq. (13c) for $\tilde{s} = s^*$. The superscript asterisk is indicative of stationary conditions, occurring after the initial fast transient period of [ES] build-up has taken place.

$$\tilde{s} = \omega(s^* \exp(s^* - e_0(\theta - \theta^*))) \tag{13b}$$

$$\tilde{\lambda} = \sqrt{(1 + e_0 + \tilde{s})^2 - 4\beta e_0} \tag{13c}$$

$$s^* = \frac{1}{2}\left(s_0 - 1 - e_0 + \sqrt{(s_0 + e_0 + 1)^2 - 4e_0 s_0}\right) \tag{13d}$$

$$\theta^* = \frac{2\beta \arctan h(\lambda^*/(1 + e_0 + s^*))}{\lambda^*} \tag{13e}$$

The choice of the stationary instant t^* is in order to simplify the usage of the PEA model given that the stationary pivotal variable $(S_0 - P^*)/v^*$ is independent of K_S. The stationary version of the PEA model can easily be used to estimate MM parameters through the application of linear regressions (Pinto et al., 2015):

$$\frac{S_0 - P^*}{v^*} = \begin{cases} \dfrac{K_M + S_0}{V_{max}}, & S_0 > E_0 \\[2ex] \dfrac{K_M + E_0}{V_{max}}, & S_0 < E_0 \end{cases} \tag{14a}$$

It should be noted that in the case of maximal dissociation constant $(K_S/K_M = 1)$, the previous equation is reduced to the

MM equation for $S_0 > E_0$, and to the simplified Bajzer and Strehler equation (Bajzer and Strehler, 2012) for $S_0 < E_0$:

$$v_0 = \begin{cases} \dfrac{V_{max}S_0}{K_M + S_0}, & S_0 > E_0 \\[2ex] \dfrac{V_{max}S_0}{K_M + E_0}, & S_0 < E_0 \end{cases} \tag{14b}$$

References

Acker, M.G., Auld, D.S., 2014. Considerations for the design and reporting of enzyme assays in high-throughput screening applications. Persp. Sci. 1, 56–73, http://dx.doi.org/10.1016/j.pisc.2013.12.001.

Baici, A., 2015. Kinetics of Enzyme-Modifier Interactions. Springer, pp. 39, http://dx.doi.org/10.1007/978-3-7091-1402-5.

Bajzer, Z., Strehler, E.E., 2012. About and beyond the Henri–Michaelis–Menten rate equation for single-substrate enzyme kinetics. Biochem. Biophys. Res. Commun. 417, 982–985, http://dx.doi.org/10.1016/j.bbrc.2011.12.051.

Berberan-Santos, M.N., 2010. A general treatment of Henri–Michaelis–Menten enzyme kinetics: exact series solution and approximate analytical solutions. MATCH Commun. Math. Comput. Chem. 63, 283–318.

Bersani, A.M., Dell'Acqua, G., 2011. Asymptotic expansions in enzyme reactions with high enzyme concentrations. Math. Methods Appl. Sci. 34, 1954–1960, http://dx.doi.org/10.1002/mma.1495.

Briggs, G.E., Haldane, J.B.S., 1925. A note on the kinetics of enzyme action. Biochem. J. 19, 338–339, http://dx.doi.org/10.1042/bj0190338.

Brown, A.J., 1902. Enzyme action. J. Chem. Soc. Trans. 81, 373–388, http://dx.doi.org/10.1039/ct9028100373.

Cornish-Bowden, A., 1987. The time dimension in steady-state kinetics: a simplified representation of control coefficients. Biochem. Educ. 15, 144–146, http://dx.doi.org/10.1016/0307-4412(87)90048-3.

Cornish-Bowden, A., 2012. Fundamentals of Enzyme Kinetics, 4th ed. Wiley-Blackwell, Weinheim, Germany, pp. 25–45, http://dx.doi.org/10.1016/b978-0-408-10617-7.50007-9, 400–410.

Dormand, J.R., Prince, P.J., 1980. A family of embedded Runge–Kutta formulae. J. Comp. Appl. Math. 6, 19–26, http://dx.doi.org/10.1016/0771-050x(80)90013-3.

Duggleby, R.G., 2001. Quantitative analysis of the time courses of enzyme-catalyzed reactions. Methods 24, 168–174, http://dx.doi.org/10.1006/meth.2001.1177.

Eisenthal, R., Danson, M.J., Hough, D.W., 2007. Catalytic efficiency and k_{cat}/K_M: a useful comparator? Trends Biotechnol. 25, 247–249, http://dx.doi.org/10.1016/j.tibtech.2007.03.010.

Fersht, A., 1999. Structure and Mechanism in Protein Science, 2nd ed. W.H. Freeman and Company, New York, pp. 110–111, 143.

Finn, N., Kemp, M., 2014. Systems biology approaches to enzyme kinetics: analyzing network models of drug metabolism. In: Nagar, S., Argikar, U.A., Tweedie, D.J. (Eds.), Enzyme Kinetics in Drug Metabolism. Humana Press, pp. 317–334, http://dx.doi.org/10.1007/978-1-62703-758-7_15.

Hanson, S.M., Schnell, S., 2008. Reactant stationary approximation in enzyme kinetics. J. Phys. Chem. A 112, 8654–8658, http://dx.doi.org/10.1021/jp8026226.

Henri, V., 1902. Théorie générale de l'action des quelques diastases. C. R. Hebd. Séances Acad. Sci. 135, 916–919, http://dx.doi.org/10.1016/j.crvi.2005.10.007.

Henri, V., 1903. Lois Générales de l'action des Diastases. Hermann, Paris, http://dx.doi.org/10.1007/bf01692307.

Johnson, K.A., Goody, R.S., 2011. The original Michaelis constant: translation of the 1913 Michaelis–Menten Paper. Biochemistry 50, 8264–8269, http://dx.doi.org/10.1021/bi201284u.

Michaelis, L., Menten, M., 1913. Die Kinetik der Invertinwirkung. Biochem. Z. 49, 333–369.

Nath, A., Atkins, W.M., 2008. A quantitative index of substrate promiscuity. Biochemistry 47, 157–166, http://dx.doi.org/10.1021/bi701448p.

Pandya, C., Farelli, J.D., Dunaway-Mariano, D., Allen, K.N., 2014. Enzyme promiscuity: engine of evolutionary innovation. J. Biol. Chem. 289, 30229–30236, http://dx.doi.org/10.1074/jbc.r114.572990.

Pinto, M.F., Estevinho, B.N., Crespo, R., Rocha, F.A., Damas, A.M., Martins, P.M., 2015. Enzyme kinetics: the whole picture reveals hidden meanings. FEBS J. 282, 2309–2316, http://dx.doi.org/10.1111/febs.13275.

Schnell, S., Maini, P.K., 2000. Enzyme kinetics at high enzyme concentration. Bull. Math. Biol. 62, 483–499, http://dx.doi.org/10.1006/bulm.1999.0163.

Segel, L.A., 1988. On the validity of the steady state assumption of enzyme kinetics. Bull. Math. Biol. 50, 579–593, http://dx.doi.org/10.1016/s0092-8240(88)80057-0.

Sols, A., Marco, R., 1970. Concentrations of metabolites and binding sites. Implications in metabolic regulation. Curr. Top. Cell. Regul. 2, 227–273, http://dx.doi.org/10.1016/b978-0-12-152802-7.50013-x.

Tzafriri, A.R., 2003. Michaelis–Menten kinetics at high enzyme concentrations. Bull. Math. Biol. 65, 1111–1129, http://dx.doi.org/10.1016/s0092-8240(03)00059-4.

Yang, J., Copeland, R.A., Lai, Z., 2009. Defining balanced conditions for inhibitor screening assays that target bisubstrate enzymes. J. Biomol. Screen. 14, 111–120, http://dx.doi.org/10.1177/1087057108328763.